WJEC
AS Mathematics

Study and Revision Guide

Stephen Doyle
edited by **Tony Holloway**

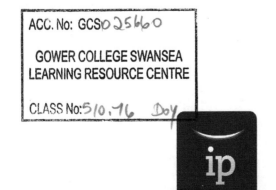

ip

Published in 2012 by Illuminate Publishing Ltd, P.O Box 1160,
Cheltenham, Gloucestershire GL50 9RW

Orders: Please visit www.illuminatepublishing.com
or email sales@illuminatepublishing.com

British Library Cataloguing in Publication Data

A catalogue record for this book is available from the British Library

ISBN 978-1-908682-02-4

Printed in England by 4edge Ltd., Hockley, Essex

04.13

The publisher's policy is to use papers that are natural, renewable and recyclable
products made from wood grown in sustainable forests. The logging and manufacturing
processes are expected to conform to the environmental regulations of the country of
origin.

This material has been endorsed by WJEC and offers high quality support for the
delivery of WJEC qualifications. While this material has been through a WJEC quality
assurance process, all responsibility for the content remains with the publisher.

Editor: Geoff Tuttle
Cover and text design: Nigel Harriss
Text and layout: The Manila Typesetting Company

Acknowledgements

I am very grateful to Rick, Geoff and the team at Illuminate Publishing for their
professionalism, support and guidance throughout this project. It has been a pleasure to
work so closely with them.

The author and publisher wish to thank:

Tony Holloway for his thorough review of the book and expert insights, observations and
contributions.

Contents

How to use this book

The contents of this study and revision guide are designed to guide you through to success in the Pure Mathematics components of the WJEC Mathematics AS level examinations. It has been written by an experienced author and teacher and edited by a senior subject expert. This book has been written specifically for the WJEC AS course you are taking and includes everything you need to know to perform well in your final exams.

Knowledge and Understanding

Topics start with a short list of the material covered in the topic and each topic will give the underpinning knowledge and skills you need to perform well in your exams.

If any formulae are included in a topic, you will be told whether you need to remember them or whether they will be given in the formula booklet.

Formulae used will be highlighted and will be included in a topic summary at the end of each unit.

The knowledge section is kept fairly short leaving plenty of space for detailed explanation of examples. Pointers will be given to the theory, examples and questions that will help you understand the thinking behind the steps. You will also be given detailed advice when it is needed.

Another feature is Grade Boost where there are tips on achieving your best grade by usually avoiding certain pitfalls which can let students down.

Exam Practice and Technique

Being able to answer examination questions, lies at the heart of this book. This means that we have included questions throughout the book that will build your skills and knowledge until you are at a stage to answer full exam questions on your own. Examples are included, some of which are based on recent examination questions. These are annotated with Pointers and general advice about the knowledge, skills and techniques needed to answer them. There is a comprehensive Q&A section in each topic that provides examination questions with commentary so you can see how the question should be answered.

There is a Test yourself section where you are encouraged to answer questions on the topic and then compare your answers with the ones given at the back of the book. You should, of course, work through complete examination papers as part of your revision process.

We advise that you look at the WJEC website www.wjec.co.uk where you can download materials such as the specification and past papers to help you with your studies. From this website you will be able to download the formula booklet which you will use in your examinations. You will also find specimen papers and mark schemes on the site.

Good luck with your revision.

Stephen Doyle   Tony Holloway

Unit C1 Pure Mathematics 1

Unit C1 covers Pure Mathematics and seeks to build on the knowledge obtained from your GCSE course. You must be proficient in the use of mathematical theories and techniques such as solving simple linear and quadratic equations, transposing formulae, algebraic manipulation, etc., and this may require you looking back over your GCSE work.

The knowledge, skills and understanding of the material in C1 will be built on in the other pure mathematics units as well as the other units in mechanics or statistics that you will study. It will be assumed when you complete these other units that you have a thorough knowledge of the material covered in C1.

Revision checklist

Tick column 1 when you have completed all the notes.
Tick column 2 when you think you have a good grasp of the topic.
Tick column 3 during the final revision when you feel you have mastery of the topic.

		1	2	3	Notes
	1 Indices and surds				
p8	Laws of indices				
p11	Use and manipulation of surds				
	2 Quadratic functions, equations, graphs and their transformations				
p15	Completing the square				
p16	Solution of quadratic equations				
p17	The discriminant of a quadratic function				
p18	Sketching a quadratic function				
p20	The solution of linear and quadratic inequalities				
p23	Transformations of the graph of $y = f(x)$				
	3 Coordinate geometry and straight lines				
p37	Finding the gradient, equation, length and mid-point of a line joining two points				
p40	Conditions for two straight lines to be parallel or perpendicular to each other				
	4 Polynomials and the binomial expansion				
p48	Algebraic manipulation of polynomials				
p54	Binomial expansion				
	5 Differentiation				
p62	Differentiation from first principles				
p63	Differentiation of x^n and related sums and differences				

		1	2	3	Notes
p65	Stationary points				
p66	The second order derivative				
p67	Increasing and decreasing functions				
p68	Simple optimisation problems				
p70	Gradients of tangents and normals, and their equations				
p73	Simple curve sketching				

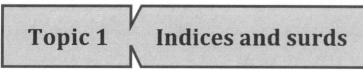

Topic 1 — Indices and surds

This topic covers the following:

- Laws of indices
- Use and manipulation of surds

Indices are another name for powers and in this topic you will learn how to manipulate powers using some basic rules.

Surds are irrational numbers; this means that they cannot be expressed as fractions, recurring decimals or terminating decimals. In this topic you will revise ways of manipulating surds.

Laws of indices

Indices are powers and, although you are unlikely to get an entire question just on them, you need to be able to use them in other areas such as differentiation and integration.

Indices are easy – all you have to do is follow some rules. All the rules apply to numbers of the same base. For example, the rules would apply to $2^5 \times 2^4$ because the bases are the same (i.e. 2). They would not apply to $2^3 \times 5^4$ where the bases are different (i.e. 2 and 5).

Multiplying with indices

You simply add the indices like this:

$$2^3 \times 2^5 = (2 \times 2 \times 2) \times (2 \times 2 \times 2 \times 2 \times 2) = 2^{3+5} = 2^8$$

Remember that: $2 = 2^1$, so

$$2 \times 2^5 \times 2^{-3} = 2^{1+5+(-3)} = 2^3$$

Generally

$$a^m \times a^n = a^{m+n}$$

Dividing with indices

You simply subtract the indices. Make sure you subtract the bottom index from the top index like this:

$$3^7 \div 3^2 = \frac{3 \times 3 \times 3 \times 3 \times 3 \times 3 \times 3}{3 \times 3} = 3^{7-2} = 3^5$$

$$\frac{2^4}{2^3} = 2^{(4-3)} = 2^1 = 2$$

$$\frac{5^5}{5^7} = 5^{(5-7)} = 5^{-2}$$

$$a^m \div a^n = a^{m-n}$$

In the third example, many students make the mistake of subtracting 5 from 7 to give positive 2. Remember that the final index does not always have to be positive.

Power raised to a power

You simply multiply the indices inside and outside the bracket like this:

$$(2^3)^5 = 2^{3 \times 5} = 2^{15}$$

$$\left(2^{\frac{1}{2}}\right)^4 = 2^{\left(\frac{1}{2} \times 4\right)} = 2^2$$

$$(2^{-2})^3 = 2^{(-2 \times 3)} = 2^{-6}$$

Generally

$$(a^m)^n = a^{m \times n} = a^{mn}$$

Negative, fractional and zero powers

A negative power means one divided by the number or letter raised to the positive power like this:

$$3^{-5} = \frac{1}{3^5}$$

$$x^{-2} = \frac{1}{x^2}$$

Generally

$$a^{-m} = \frac{1}{a^m} \qquad \text{(provided } a \neq 0)$$

Fractional powers mean roots. If the denominator (i.e. bottom number) is a 2 then it is a square root and if it is a 3 then it is a cube root. For example:

$$4^{\frac{1}{2}} = \sqrt{4} = 2$$

$$8^{\frac{1}{3}} = \sqrt[3]{8} = 2$$

Where there is a fraction with a numerator (i.e. top number) larger than 1, the number inside the root is raised to the power of the numerator.

In the following example x is raised to the power $\frac{2}{3}$. The denominator (i.e. 3) means the cube root of x and the numerator (i.e. 2) means that the x is squared inside the root. It does not matter whether the cube rooting or the squaring is done first.

$$x^{\frac{2}{3}} = \sqrt[3]{x^2} = \left(\sqrt[3]{x}\right)^2$$

$$8^{\frac{2}{3}} = \sqrt[3]{8^2} = \sqrt[3]{64} = 4$$

or

$$8^{\frac{2}{3}} = \left(\sqrt[3]{8}\right)^2 = 2^2 = 4$$

> If you have a number raised to a fractional power, you need to change it to roots and powers as shown here. It does not matter whether you perform the root first and then raise the answer to the power or the other way around. Do whichever is easiest. As a general rule, finding the root first will keep the numbers low and therefore more recognisable.

> It is easier to find the cube root first and then square the answer as shown here.

Generally

$$a^{\frac{m}{n}} = \sqrt[n]{a^m} = \left(\sqrt[n]{a}\right)^m$$

Combinations of fractional and negative powers

A negative power means the reciprocal (i.e. 1 over the number raised to the positive power) and the fraction means a root.

$$27^{-\frac{1}{3}} = \frac{1}{27^{\frac{1}{3}}} = \frac{1}{\sqrt[3]{27}} = \frac{1}{3}$$

$$x^{-\frac{3}{2}} = \frac{1}{\sqrt{x^3}} \text{ or } \frac{1}{(\sqrt{x})^3}$$

$$16^{-\frac{3}{2}} = \frac{1}{(\sqrt{16})^3} = \frac{1}{4^3} = \frac{1}{64}$$

Generally

Here you have to find $(\sqrt{16})^3$. Because 16 is a perfect square, it is easier to find the square root of 16 and then cube the answer rather than cube 16 and then have to square root the answer.

$$a^{-\frac{m}{n}} = \frac{1}{a^{\frac{m}{n}}} = \frac{1}{\sqrt[n]{a^m}} \text{ or } \frac{1}{(\sqrt[n]{a})^m}$$

Zero powers

Any number raised to a zero power is always 1, even if you do not know what the number is. For example $x^0 = 1$ or $(ab)^0 = 1$.

$$3^0 = 1$$

$$0.5^0 = 1$$

Generally

If $a \neq 0$,

$$a^0 = 1$$

Example

① Write the following equation using indices:

$$y = \frac{3}{4}\sqrt[3]{x} - \frac{6}{x^2} + 1$$

Answer

$$y = \frac{3}{4}x^{\frac{1}{3}} - 6x^{-2} + 1$$

Example

② Given that $y = 8x^{-2} + \dfrac{3}{2}x^{-\frac{1}{2}}$, find y when $x = 4$.

Answer

Notice that if you are substituting numbers into the above equation, you need to change from index form into roots, etc. This makes it easier to substitute the numbers in.

$$y = \frac{8}{x^2} + \frac{3}{2\sqrt{x}} = \frac{8}{4^2} + \frac{3}{2\sqrt{4}} = \frac{1}{2} + \frac{3}{4} = 1\frac{1}{4}$$

Use and manipulation of surds

Numbers like $\sqrt{18}$ are called surds. Surds are irrational numbers. This means that they cannot be expressed as fractions, recurring decimals or terminating decimals. Surds can be simplified like this:

$$\sqrt{18} = \sqrt{9 \times 2} = 3\sqrt{2}$$

Here the number 18 is written as the product of two factors that include a square number. 9 is a perfect square so can be square-rooted to give a whole number answer. So $\sqrt{18} = 3\sqrt{2}$.

> Always try to find the largest square factor. For example $\sqrt{80}$ could be written as $\sqrt{16 \times 5} = 4\sqrt{5}$ or $\sqrt{4 \times 20}$ but this still needs further simplification to $\sqrt{4 \times 4 \times 5} = 4\sqrt{5}$. It is quicker to spot that 16 is the highest square factor of 80, so we have $\sqrt{80} = \sqrt{16 \times 5} = 4\sqrt{5}$.

Simplifying surds

Here are some general rules when manipulating surds:

$$\sqrt{a} \times \sqrt{a} = a$$

$$\sqrt{a} \times \sqrt{b} = \sqrt{ab}$$

$$\left(\sqrt{a} + \sqrt{b}\right)\left(\sqrt{a} - \sqrt{b}\right) = a - b$$

The following examples show ways in which surds can be simplified:

1   $\left(\sqrt{3}\right)^2 = \sqrt{3} \times \sqrt{3} = 3$

2   $\left(5\sqrt{2}\right)^2 = 5\sqrt{2} \times 5\sqrt{2} = 25 \times 2 = 50$

3   $\left(3\sqrt{2}\right) \times \left(4\sqrt{2}\right) = 12 \times 2 = 24$

4   $3\sqrt{2} + 2\sqrt{2} = 5\sqrt{2}$

5   $\left(2 + \sqrt{7}\right)\left(2 + \sqrt{7}\right) = 2\left(2 + \sqrt{7}\right) + \sqrt{7}\left(2 + \sqrt{7}\right) = 4 + 2\sqrt{7} + 2\sqrt{7} + 7 = 11 + 4\sqrt{7}$

6   $(1+\sqrt{3})(5-\sqrt{12})=1(5-\sqrt{12})+\sqrt{3}(5-\sqrt{12})$

$$=5-\sqrt{12}+5\sqrt{3}-\sqrt{3\times12}$$

$$=5-2\sqrt{3}+5\sqrt{3}-\sqrt{36}$$

$$=-1+3\sqrt{3}$$

Rationalising surds

If you have a fraction with a surd on the bottom, then it needs to be removed (i.e. rationalised). This is done by multiplying the top (i.e. numerator) and bottom (i.e. denominator) of the fraction by the surd. Rationalising makes sure that the denominator is no longer an irrational number.

$\dfrac{1}{\sqrt{3}}=\dfrac{1}{\sqrt{3}}\times\dfrac{\sqrt{3}}{\sqrt{3}}=\dfrac{\sqrt{3}}{3}$ The fraction is simplified when there are no surds in the denominator.

When there is a fraction containing a denominator like this $\dfrac{1}{1-\sqrt{2}}$, to remove the irrational number in the denominator, both the numerator (i.e. top) and denominator (i.e. bottom) of the fraction are multiplied by the conjugate of the denominator which in this case is $1+\sqrt{2}$. The conjugate is the same as the denominator except the sign is the opposite.

Hence $\dfrac{1}{1-\sqrt{2}}=\dfrac{1}{(1-\sqrt{2})}\times\dfrac{(1+\sqrt{2})}{(1+\sqrt{2})}=\dfrac{1+\sqrt{2}}{1-2}=\dfrac{1+\sqrt{2}}{-1}=-1-\sqrt{2}$

Example

In both of these questions you have been asked to simplify. In each case this is done by rationalising the denominator (i.e. by removing the surds from the denominator) and simplifying the result.

①   Simplify

$\dfrac{10}{\sqrt{5}}$

Answer

$\dfrac{10}{\sqrt{5}}=\dfrac{10}{\sqrt{5}}\times\dfrac{\sqrt{5}}{\sqrt{5}}=\dfrac{10\sqrt{5}}{5}=2\sqrt{5}$

Example

②   Simplify

$\dfrac{1}{2-\sqrt{5}}$

> Rationalise the denominator by multiplying the numerator and denominator by the conjugate of the denominator.

Answer

$\dfrac{1}{(2-\sqrt{5})}\dfrac{(2+\sqrt{5})}{(2+\sqrt{5})}=\dfrac{2+\sqrt{5}}{4-5}=\dfrac{2+\sqrt{5}}{-1}=-2-\sqrt{5}$

Examination style questions

① Simplify

$\sqrt{45} + \sqrt{80} + \sqrt{125}$ [3]

> Remember to spot those factors that are perfect squares.

Answer

$\sqrt{45} + \sqrt{80} + \sqrt{125} = \sqrt{9 \times 5} + \sqrt{16 \times 5} + \sqrt{25 \times 5} = 3\sqrt{5} + 4\sqrt{5} + 5\sqrt{5} = 12\sqrt{5}$

② Simplify

$\dfrac{3\sqrt{3} - \sqrt{2}}{\sqrt{3} - \sqrt{2}}$ [4]

Answer

$\dfrac{3\sqrt{3} - \sqrt{2}}{\sqrt{3} - \sqrt{2}} = \dfrac{(3\sqrt{3} - \sqrt{2})(\sqrt{3} + \sqrt{2})}{(\sqrt{3} - \sqrt{2})(\sqrt{3} + \sqrt{2})} = \dfrac{9 + 3\sqrt{6} - \sqrt{6} - 2}{3 + \sqrt{6} - \sqrt{6} - 2} = \dfrac{7 + 2\sqrt{6}}{1} = 7 + 2\sqrt{6}$

③ Simplify

$\dfrac{3}{\sqrt{3}} + \sqrt{75} + (\sqrt{2} \times \sqrt{6})$ [4]

Answer

$\dfrac{3}{\sqrt{3}} + \sqrt{75} + (\sqrt{2} \times \sqrt{6}) = \dfrac{\sqrt{3 \times 3}}{\sqrt{3}} + \sqrt{25 \times 3} + (\sqrt{2} \times \sqrt{2 \times 3}) = \sqrt{3} + 5\sqrt{3} + 2\sqrt{3} = 8\sqrt{3}$

Test yourself

Answer the following questions and check your answers before moving on to the next topic.

① Write the following equation using indices:

$y = 5\sqrt{x} + \dfrac{45}{x} - 7$

② Simplify each of the following without using a calculator:

   (a) 5^0

   (b) 3^{-2}

   (c) $8^{\frac{1}{3}}$

   (d) $25^{-\frac{1}{2}}$

   (e) $16^{\frac{3}{2}}$

③ Simplify each of the following, expressing your answers in surd form.

   (a) $\sqrt{48} + \dfrac{12}{\sqrt{3}} - \sqrt{27}$

   (b) $\dfrac{2 + \sqrt{5}}{3 + \sqrt{5}}$

(Note: answers to Test yourself are found at the back of the book.)

1	Simplify	
(a)	$\dfrac{5\sqrt{7}-\sqrt{3}}{\sqrt{7}-\sqrt{3}}$	[4]
(b)	$\left(\sqrt{15}\times\sqrt{20}\right)-\sqrt{75}-\dfrac{\sqrt{60}}{\sqrt{5}}$	[4]

(WJEC C1 May 2010 Q2)

Answer

1 (a)

$$\frac{5\sqrt{7}-\sqrt{3}}{\sqrt{7}-\sqrt{3}}=\frac{\left(5\sqrt{7}-\sqrt{3}\right)\left(\sqrt{7}+\sqrt{3}\right)}{\left(\sqrt{7}-\sqrt{3}\right)\left(\sqrt{7}+\sqrt{3}\right)}$$

Remember you have to multiply top and bottom by the conjugate of the bottom.

$$=\frac{5\times7+5\sqrt{7}\sqrt{3}-\sqrt{3}\sqrt{7}-3}{7+\sqrt{7}\sqrt{3}-\sqrt{3}\sqrt{7}-3}$$

Remember that $\sqrt{7}\sqrt{3}=\sqrt{3}\sqrt{7}=\sqrt{21}$ so the middle two terms can be subtracted.

$$=\frac{35+4\sqrt{7}\sqrt{3}-3}{4}=\frac{32+4\sqrt{7}\sqrt{3}}{4}$$

Notice that the 4 on the bottom can be divided into both terms on the top.

$$=8+\sqrt{21}$$

(b)

$$\left(\sqrt{15}\times\sqrt{20}\right)-\sqrt{75}-\frac{\sqrt{60}}{\sqrt{5}}$$

60 is split into the factors 5 and 12. These are chosen as there is a $\sqrt{5}$ on the bottom which will cancel.

$$=\sqrt{300}-\sqrt{3\times25}-\frac{\sqrt{5\times12}}{\sqrt{5}}$$

$\sqrt{5}$ is cancelled in the fraction.

$$=\sqrt{3\times100}-\sqrt{3\times25}-\sqrt{12}$$

The numbers inside the roots are written as the product of two factors where one of the factors is a perfect square.

$$=10\sqrt{3}-5\sqrt{3}-\sqrt{4\times3}$$

$$=10\sqrt{3}-5\sqrt{3}-2\sqrt{3}$$

$$=3\sqrt{3}$$

<table>
<tr><td>

Topic 2

</td><td>

Quadratic functions, equations, graphs and their transformations

</td></tr>
</table>

This topic covers the following:

- Completing the square
- Solution of quadratic equations
- The discriminant of a quadratic function
- Quadratic functions and their graphs
- The solution of linear and quadratic inequalities
- Transformations of the graph of $y = f(x)$

Completing the square

A quadratic expression $ax^2 + bx + c$ can be written in the form $(x + p)^2 + q$ and this is called completing the square. Note that the values of p and q can be positive or negative.

For example, suppose you were asked to express $x^2 + 6x + 11$ in the form $(x + a)^2 + b$, where the values of a and b are to be determined.

Provided there is no number other than 1 in front of the x^2 (called the coefficient of x^2), halve the number in front of the x (the coefficient of x). Here there is a 6 in front of the x so halving this gives 3. If there is a minus sign, then this will need to be included.

The number including the sign is substituted into a bracket like this:

$(x + 3)^2$

When this is expanded it gives $x^2 + 6x + 9$. So we have the first two terms and also a number 9 which we remove by subtracting outside the bracket like this:

$(x + 3)^2 - 9$

It is now necessary to add the 11, so we have

$x^2 + 6x + 11 = (x + 3)^2 - 9 + 11 = (x + 3)^2 + 2$

The answer can be compared with $(x + a)^2 + b$

Hence $a = 3$ and $b = 2$.

When the coefficient of x^2 is not equal to 1, we use the following method:

Example

① Express $2x^2 + 12x + 3$ in the form $a(x + b)^2 + c$, where a, b and c are to be determined.

Answer

① $2x^2 + 12x + 3$

$= 2\left[x^2 + 6x + \dfrac{3}{2}\right]$

$= 2\left[(x + 3)^2 - 9 + \dfrac{3}{2}\right]$

> Before completing the square, take 2 out as a factor because the coefficient of x^2 needs to be one when completing the square.

> We now complete the square inside the square bracket.

$$= 2\left[(x+3)^2 - \frac{15}{2}\right]$$

$$= 2(x+3)^2 - 15$$

Hence $a = 2, b = 3, c = -15$

> Now multiply by the two outside the square bracket to give the required format.

Solution of quadratic equations

Quadratic equations are equations that can be written in the form: $ax^2 + bx + c = 0$

There are three ways to solve quadratic equations:

1 By factorising. You should be familiar with this from your GCSE work.
2 By completing the square.
3 By using the quadratic formula.

Solving a quadratic equation by factorising

Example

① Solve $2x^2 + 7x - 4 = 0$

Answer

$(2x - 1)(x + 4) = 0$

Substituting each bracket equal to 0 gives $2x - 1 = 0$ or $x + 4 = 0$

Hence $x = \dfrac{1}{2}$ or $x = -4$

Grade boost

Make sure you can factorise quadratic expressions and then solve quadratic equations as you will need this knowledge in many parts of the AS and A2 course.

Solving a quadratic equation by completing the square

Example

① Show that $x^2 + 0.8x - 3.84$ may be expressed in the form $(x + p)^2 - 4$, where p is a constant whose value is to be found.

Hence solve the quadratic equation $x^2 + 0.8x - 3.84 = 0$.

Answer

$$x^2 + 0.8x - 3.84 = (x + 0.4)^2 - 0.16 - 3.84$$
$$= (x + 0.4)^2 - 4$$

Hence $p = 0.4$

$x^2 + 0.8x - 3.84 = 0$

So, $(x + 0.4)^2 - 4 = 0$

$(x + 0.4)^2 = 4$

Grade boost

You must use completing the square to solve the quadratic equation. You will lose marks if a method is specified in the question and you use a different method.

Square-rooting both sides gives:

$(x + 0.4) = \pm 2$

So, $x = 2 - 0.4$ or $x = -2 - 0.4$

Hence $x = 1.6$ or $x = -2.4$

> You must include both the positive and negative values when you square-root a number.

Solving a quadratic equation using the formula

Quadratic equations, when in the form $ax^2 + bx + c = 0$ can be solved using the formula:

$$x = \frac{-b \pm \sqrt{b^2 - 4ac}}{2a}$$

> Be careful with signs when you are entering numbers into this equation.

> Calculators are not allowed in the Core 1 examination. The numbers you will have to enter into the quadratic formula will be relatively simple numbers.

Important note: This formula will not be given in the formula booklet so you will need to remember it.

Example

① Solve the equation $2x^2 - x - 6 = 0$

Answer

① Comparing the equation given, with $ax^2 + bx + c = 0$ gives $a = 2$, $b = -1$ and $c = -6$.

Substituting these values into the quadratic equation formula gives:

$$x = \frac{1 \pm \sqrt{(-1)^2 - 4(2)(-6)}}{2(2)}$$

> If you get asked to leave your answer in surd form, then you must use either the formula or the method involving completing the square.

$$= \frac{1 \pm \sqrt{1 + 48}}{4} = \frac{1 \pm 7}{4} = \frac{1+7}{4} \text{ or } \frac{1-7}{4} = 2 \text{ or } -1.5$$

The discriminant of a quadratic function

The roots of a quadratic equation are the same as the solutions and are also the x-coordinates of the points where the graph of the equation cuts the x-axis.

$ax^2 + bx + c$ is a quadratic function. The quantity $b^2 - 4ac$ is called the discriminant and it gives the following information about the roots of the quadratic equation $ax^2 + bx + c = 0$:

If $b^2 - 4ac > 0$, then there are two real and distinct (i.e. different) roots.

If $b^2 - 4ac = 0$, then there are two real and equal roots.

If $b^2 - 4ac < 0$, then there are no real roots.

The three situations can be shown graphically (for $a > 0$):

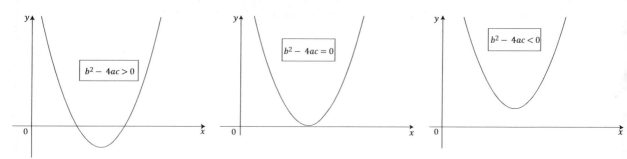

Sketching a quadratic function

The quadratic equation having the form $y = ax^2 + bx + c$ has a graph which is a parabola.

Depending on the sign of a in the above equation, the parabola is $\cup$-shaped if a is positive or $\cap$-shaped if a is negative.

To find the points where the parabola intersects the x-axis, you can solve the equation $ax^2 + bx + c = 0$.

If a question starts by asking you to complete the square and then later in the question asks you to sketch the curve and/or find the maximum or minimum value then there is a quick way of doing this.

When the square has been completed, the equation for the curve will be in this format:

$y = a(x + p)^2 + q$

When $x = -p$, the value of the bracket is zero and, since the bracket is squared, this is its minimum value (since it cannot be negative), hence the minimum value of y is q.

From the equation $y = a(x + p)^2 + q$

If a is positive (i.e. $a > 0$) the curve will be $\cup$-shaped.

If a is negative (i.e. $a < 0$) the curve will be $\cap$-shaped.

The vertex (i.e. the maximum or minimum point) will be at $(-p, q)$.

The axis of symmetry will be $x = -p$

For example the curve with the equation $y = 2(x + 3)^2 - 1$ can be compared with $y = a(x + p)^2 + q$. This gives $a = 2, p = 3$ and $q = -1$.

The curve will be $\cup$-shaped with a minimum point at $(-3, -1)$ (i.e. $(-p, q)$) and an axis of symmetry of $x = -3$.

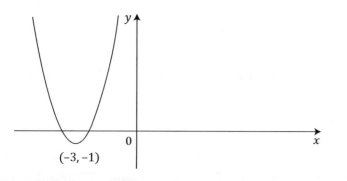

$(-3, -1)$

Example

① Express $3x^2 - 12x + 17$ in the form $a(x + b)^2 + c$, where the values of the constants a, b and c are to be found.

Hence, sketch the graph of $y = 3x^2 - 12x + 17$, indicating the coordinates of its stationary point.   [5]

(WJEC C1 Jan 09 Q4)

Answer

① $3x^2 - 12x + 17$

> The 3 must be taken out as a factor first before completing the square.

$$= 3\left[x^2 - 4x + \frac{17}{3}\right]$$

$$= 3\left[(x - 2)^2 - 4 + \frac{17}{3}\right]$$

$$= 3\left[(x - 2)^2 + \frac{5}{3}\right]$$

> Compare your answer with the format for the expression in the question which in this case is $a(x + b)^2 + c$ to find the values of a, b and c.

$$= 3(x - 2)^2 + 5$$

Hence, $a = 3$, $b = -2$ and $c = 5$

$$y = 3x^2 - 12x + 17$$

Using your answer from completing the square gives:

$$y = 3(x - 2)^2 + 5$$

This equation is in the format $y = a(x + p)^2 + q$ where $a = 3$ (which is positive so the graph will be ∪-shaped). When $x = 2$, the value of the bracket is zero, so the minimum value of y is 5. The vertex (a minimum point in this case) will be at $(-p, q)$ which gives the point $(2, 5)$.

Although you are not asked to find where the curve cuts the y-axis, finding this point is useful when drawing the graph. In other questions you may be specifically asked to find the y-coordinate of this point.

To find where the curve cuts the y-axis, substitute $x = 0$ into the equation for the curve $y = 3x^2 - 12x + 17$ giving $y = 17$.

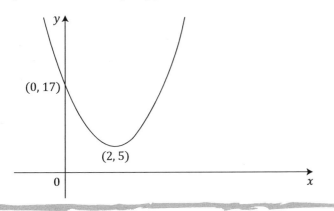

The solution of linear and quadratic inequalities

Solving linear inequalities

These are solved in a similar way to solving ordinary linear equations but there is one important difference. If you multiply or divide both sides by a negative quantity, then the inequality sign must be reversed since for example, $3 > 2$, but $-3 < -2$.

Example

① Solve the inequality $3x - 7 < 2$

Answer

① $3x - 7 < 2$

$3x < 9$ (Adding 7 to both sides)

$x < 3$ (Dividing both sides by 3)

Example

② Solve the inequality $1 - 2x > 5$

Answer

② $1 - 2x > 5$

$-2x > 4$ (Subtracting 1 from both sides)

$x < -2$ (Dividing both sides by -2 and reversing the sign)

Solving quadratic inequalities

Example

① Find the range of values of x satisfying the inequality

$2x^2 + x - 6 \leq 0$

Answer

① $2x^2 + x - 6 \leq 0$

Considering the case where $2x^2 + x - 6 = 0$ and factorising gives

$(2x - 3)(x + 2) = 0$

Giving the critical values $x = \dfrac{3}{2}$ or $x = -2$ (these are the points where the curve cuts the x-axis)

Sketching the curve for $y = 2x^2 + x - 6$ gives the following:

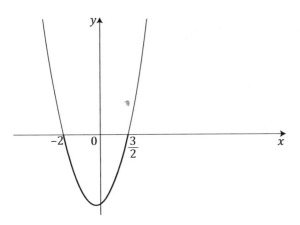

We want the part of the graph which is on or below the x-axis because of the $\leq$ in the inequality.

The range of values of x for which this occurs is $-2 \leq x \leq \dfrac{3}{2}$

Example

② Find the range of values of x satisfying the inequality $3x^2 + 2x - 1 > 0$

Answer

② $3x^2 + 2x - 1 > 0$

Considering the case where $3x^2 + 2x - 1 = 0$ and factorising gives

$(3x - 1)(x + 1) = 0$

Giving the critical valules $x = \dfrac{1}{3}$ or $x = -1$ (these are the points where the curve cuts the x-axis)

Sketching the curve for $y = 3x^2 + 2x - 1$ gives the following:

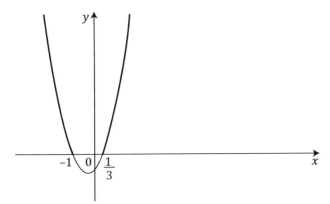

We want the part of the graph which is above the x-axis because of the $>$ in the inequality.

The range of values of x for which this occurs is

$x < -1$ or $x > \dfrac{1}{3}$

Important note: The value of x that satisfies the inequality is *either* less than -1 *or* greater than $\dfrac{1}{3}$. If you write 'and' instead 'or', your may lose a mark.

Simultaneous equations

In GCSE you solved two linear equations simultaneously to find x and y values fitting both equations. You were finding the coordinates of the point of intersection of two straight lines.

In Core 1 you need to find the solution of one linear equation and one quadratic equation. Here you will be finding the points of intersection or the point of contact of a straight line and a curve.

Example

①   Solve the simultaneous equations $y = 10x^2 - 5x - 2$ and $y = 2x - 3$ algebraically. Write down a geometrical interpretation of your results.

Answer

①   Equating expressions for y gives

$$10x^2 - 5x - 2 = 2x - 3$$

$$10x^2 - 7x + 1 = 0$$

> At the points of intersection, the y-coordinates of the curve and straight line will be the same.

Factorising this quadratic gives

$$(5x - 1)(2x - 1) = 0$$

Hence $x = \dfrac{1}{5}$ or $x = \dfrac{1}{2}$

Substituting $x = \dfrac{1}{5}$ into $y = 2x - 3$ gives

> It is easier to substitute the x-coordinate into the equation of the straight line rather than the curve.

$$y = -2\dfrac{3}{5}$$

Substituting $x = \dfrac{1}{2}$ into $y = 2x - 3$ gives

$$y = -2$$

There are two places where the line and curve intersect.

The points of intersection of the line with the curve are $\left(\dfrac{1}{5}, -2\dfrac{3}{5}\right)$ and $\left(\dfrac{1}{2}, -2\right)$

Transformations of the graph of $y = f(x)$

If you are given a graph of a function in the form $y = f(x)$, then the graph of a new function may be obtained from the original graph by applying a simple transformation.

The simple transformations include reflections, translations and stretches.

$y = f(x)$ to $y = f(x + a)$

This represents a translation of $-a$ units parallel to the x-axis. This can be represented by the translation $\begin{pmatrix} -a \\ 0 \end{pmatrix}$. Note that if a is positive in the new function, the graph moves a units to the left and if it is negative it will move a units to the right.

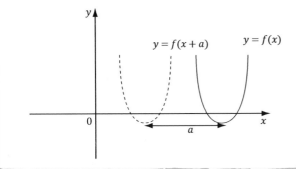

$y = f(x)$ to $y = f(x) + a$

This represents a translation of a units parallel to the y-axis. This can be represented by the translation $\begin{pmatrix} 0 \\ a \end{pmatrix}$. If a is positive, the whole graph moves up a units and if a is negative it moves down by a units.

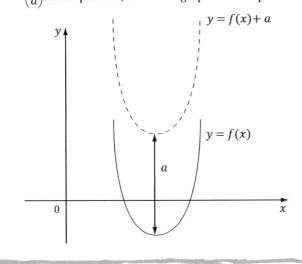

$y = f(x)$ to $y = af(x)$

This represents a one-way stretch with scale factor a parallel to the y-axis. This means that the y-value of any point on the curve will be multiplied by a leaving the x-value unchanged. It is important to note that any points of intersection with the x-axis will remain unchanged during the stretch.

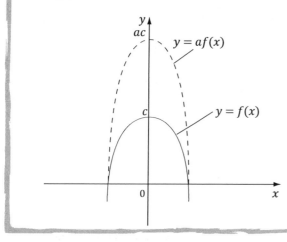

Note that if a is negative the curve will be reflected in the x-axis

$y = f(x)$ to $y = f(ax)$

This represents a one-way stretch with scale factor $\dfrac{1}{a}$ parallel to the x-axis.

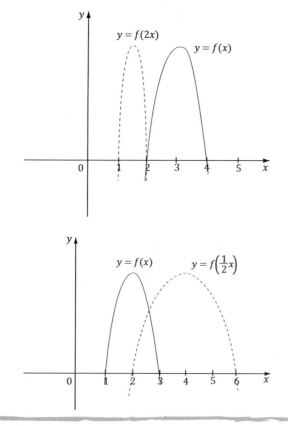

Notice the way each x-value on the graph for the original function is halved. Here you can see that the graph of $y = f(x)$ cuts the x-axis at $x = 2$ and $x = 4$. The graph of $y = f(2x)$ cuts the x-axis at half of these values i.e. $x = 1$ and $x = 2$.

Here the x-values for the original function are doubled. The graph of $y = f(x)$ cuts the x-axis at $x = 1$ and $x = 3$. The graph of $y = f\left(\dfrac{1}{2}x\right)$ cuts the x-axis at double these values i.e. $x = 2$ and $x = 6$.

Example

① The diagram shows a sketch of the graph $y = f(x)$. The graph passes through the points $(1, 0)$ and $(5, 0)$ and has a minimum point at $(3, -4)$.

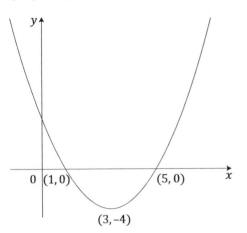

Sketch the following graphs, using a separate set of axes for each graph. In each case, you should indicate the coordinates of the stationary point and the coordinates of the points of intersection of the graph with the x-axis.

(a) $y = f(x + 1)$

(b) $y = -2f(x)$

Answer

(a) $y = f(x + 1)$ is a translation of $y = f(x)$ by one unit to the left parallel to the x-axis, i.e. a translation of $\begin{pmatrix} -1 \\ 0 \end{pmatrix}$.

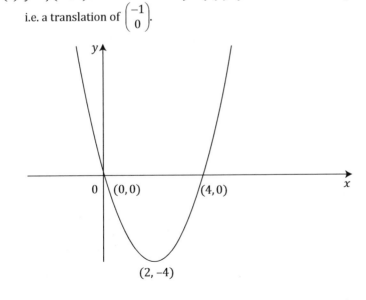

(b) $y = -2f(x)$ is a reflection in the x-axis (because of the negative sign) followed by a stretch parallel to the y-axis with scale factor 2. Note it does not matter in which order these two transformations are applied to the original function.

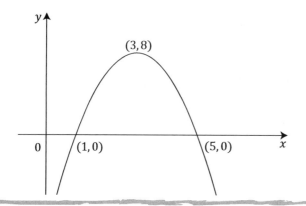

Example

② Figure 1 shows a sketch of the graph of $y = f(x)$. The graph has a maximum point at $(2, 5)$ and intersects the x-axis at the points $(-2, 0)$ and $(6, 0)$.

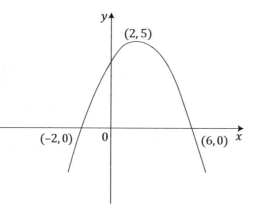

Figure 1

(a) Sketch the graph of $y = f\left(\dfrac{x}{2}\right)$, indicating the coordinates of the stationary point and the coordinates of the points of intersection of the graph with the x-axis. [3]

(b) Figure 2 shows a sketch of the graph having **one** of the following equations with an appropriate value of either p, q or r.

$y = f(x + p)$, where p is a constant

$y = f(x) + q$, where q is a constant

$y = rf(x)$, where r is a constant

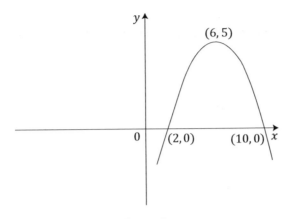

Figure 2

Write down the equation of the graph sketched in Figure 2, together with the value of the corresponding constant. [2]

(WJEC C1 Jan 2010 Q9)

Answer

② (a)

The transformation from $y = f(x)$ to $y = f(ax)$ represents a one-way stretch with scale factor $\dfrac{1}{a}$ parallel to the x-axis.

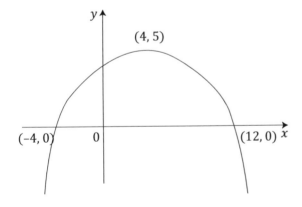

Here the transformation is from $y = f(x)$ to $y = f\left(\dfrac{x}{2}\right)$. This represents a one-way stretch with scale factor 2 parallel to the x-axis. Each x-coordinate is multiplied by 2 but the y-coordinates remain the same.

(b)

$y = f(x + p)$ represents a translation of $\begin{pmatrix} -p \\ 0 \end{pmatrix}$

$y = f(x) + q$, represents a translation of $\begin{pmatrix} 0 \\ q \end{pmatrix}$

$y = rf(x)$, represents a one-way stretch with scale factor r parallel to the y-axis

By observing the two graphs, you can see that in the transformed graph the y-coordinates have stayed the same but all the x-coordinates have moved 4 units to the right. This is a translation of $\begin{pmatrix} 4 \\ 0 \end{pmatrix}$

The equation of the translated curve is $y = f(x - 4)$

Examination style questions

① Show that $x^2 - 1.2x - 3.64$ may be expressed in the form $(x + p)^2 - 4$, where p is a constant whose value is to be found.

Hence solve the quadratic equation $x^2 - 1.2x - 3.64 = 0$. [5]

Answer

① $x^2 - 1.2x - 3.64 = (x - 0.6)^2 - 0.36 - 3.64$

$\qquad\qquad\qquad\quad = (x - 0.6)^2 - 4$

> This is obtained by completing the square.

Hence $p = -0.6$

$x^2 - 1.2x - 3.64 = 0$

So $(x - 0.6)^2 - 4 = 0$

$\qquad (x - 0.6)^2 = 4$

$\qquad x - 0.6 = \pm 2$

$\qquad x = 2 + 0.6$ or $x = -2 + 0.6$

Hence $x = 2.6$ or -1.4

② Given that $k \neq 1$ the following quadratic equation

$(k - 1)x^2 + kx + k = 0$

has two distinct real roots, show that

$3k^2 - 4k < 0$

Find the range of values of k satisfying this inequality. [5]

Answer

② For distinct and real roots

$b^2 - 4ac > 0$

Hence $k^2 - 4(k - 1)(k) > 0$

$\qquad k^2 - 4k^2 + 4k > 0$

$\qquad -3k^2 + 4k > 0$

$\qquad 3k^2 - 4k < 0$

> Remember to reverse the inequality when dividing by -1.

Factorising gives $k(3k - 4) < 0$

If a graph of $y = k(3k - 4)$ was plotted with values of k on the x-axis, as there is a positive

coefficient of k^2 the curve will be $\cup$-shaped cutting the x-axis at $k = 0$ and $k = \dfrac{4}{3}$.

Without drawing the graph you can see that the section of the graph needed will be below the x-axis.

Hence the required range of k is

$$0 < k < \frac{4}{3}$$

③ Express $4x^2 - 12x + 9$ in the form $a(x + b)^2 + c$ where the values of b and c are to be determined. [4]

Hence, sketch the graph of $4x^2 - 12x + 9$, including the coordinates of the stationary point. [3]

Answer

③ $4x^2 - 12x + 9 = 4\left[x^2 - 3x + \frac{9}{4}\right]$

$$= 4\left[\left(x - \frac{3}{2}\right)^2 - \frac{9}{4} + \frac{9}{4}\right]$$

$$= 4\left(x - \frac{3}{2}\right)^2$$

Comparing the above expression with $a(x + b)^2 + c$, gives $a = 4$, $b = -\frac{3}{2}$ and $c = 0$.

The graph of $4\left(x - \frac{3}{2}\right)^2$ is $\cup$-shaped because the coefficient of x^2 is positive.

$y = 4\left(x - \frac{3}{2}\right)^2$ has its minimum point when $x = \frac{3}{2}$. When $x = \frac{3}{2}$, $y = 0$.

Hence the coordinates of the stationary point are $\left(\frac{3}{2}, 0\right)$

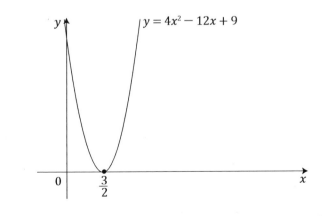

④ Solve the inequality $x^2 - 2x - 15 \leq 0$ [3]

Answer

④ Factorising $x^2 - 2x - 15 = 0$ gives

$(x - 5)(x + 3) = 0$

Hence $x = 5$ or -3

As the coefficient of x^2 is positive the graph of $x^2 - 2x - 15$ is $\cup$-shaped.

Now $x^2 - 2x - 15 \leq 0$. This is the region below the x-axis (i.e. where $y \leq 0$).

Hence $-3 \leq x \leq 5$

---

⑤ The diagram shows the graph of $y = f(x)$. The graph has a maximum point at $(1, 2)$.

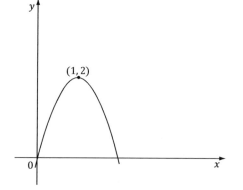

Sketch the following graphs, using a separate set of axes for each graph marking on your graphs the coordinates of the stationary point in each case:

(a) $y = -f(x)$ [2]

(b) $y = 3f(x)$ [2]

(c) $y = f(x - 1)$ [2]

(d) $y = f(2x)$ [2]

Answer

⑤ (a) $y = -f(x)$

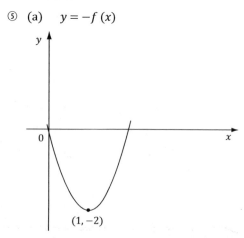

> This transformation represents a reflection in the x-axis.

(b) $y = 3f(x)$

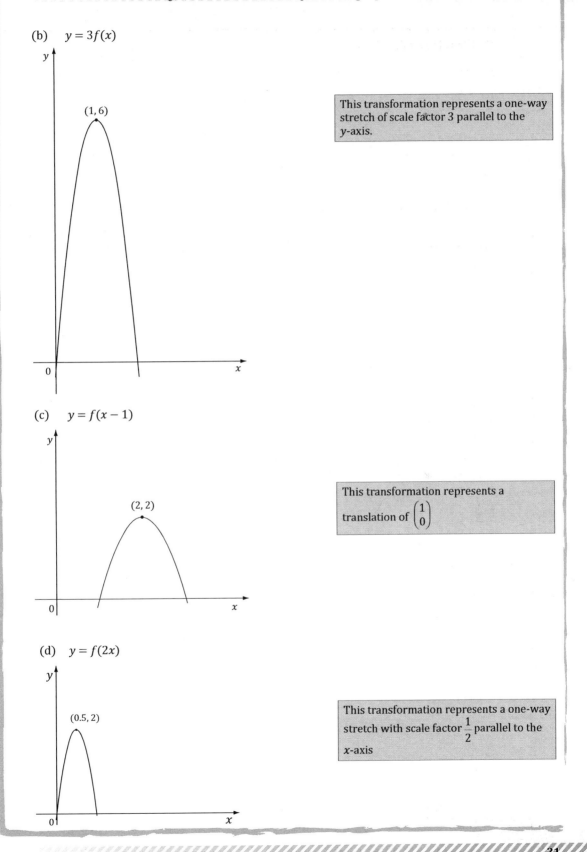

This transformation represents a one-way stretch of scale factor 3 parallel to the y-axis.

(c) $y = f(x - 1)$

This transformation represents a translation of $\begin{pmatrix} 1 \\ 0 \end{pmatrix}$

(d) $y = f(2x)$

This transformation represents a one-way stretch with scale factor $\dfrac{1}{2}$ parallel to the x-axis

Test yourself

Answer the following questions and check your answers before moving onto the next topic.

① Find the range of values of k for which the quadratic equation

$$kx^2 + 5x - 7 = 0$$

has no real roots.

② Solve the inequality $x^2 - 6x + 8 > 0$

③ Express $5x^2 - 20x + 10$ in the form $a(x + b)^2 + c$, where a, b and c are constants whose values are to be found.

④ Solve the inequality $1 - 3x < x + 7$

⑤ Show that the straight line $y = x + 4$ touches the curve $y = x^2 - 7x + 20$ and find the coordinates of the point of contact.

(Note: answers to Test yourself are found at the back of the book.)

Q & A

1

1 (a) Express $x^2 + 6x - 4$ in the form $(x + a)^2 + b$ where the values of a, b are to be determined. [2]

(b) Use your results to part (a) to find the least value of $2x^2 + 12x - 8$ and the corresponding value of x. [2]

(WJEC C1 May 2008 Q5)

Answer

1(a) $x^2 + 6x - 4 = (x + 3)^2 - 9 - 4$

$\qquad = (x + 3)^2 - 13$

Hence $a = 3$ and $b = -13$

(b) $2x^2 + 12x - 8 = 2(x^2 + 6x - 4)$

Notice that the expression in the bracket is the same as in part (a).

Using the answer to part (a) gives $2[(x + 3)^2 - 13]$

Multiplying the contents of the square bracket by the two outside gives

$2(x + 3)^2 - 26$

Now the expression is in the form $a(x + p)^2 + q$ where $a = 2$ (which is positive so the graph will be $\cup$-shaped). Also, the vertex (a minimum point in this case) will be at $(-p, q)$ which gives the point $(-3, -26)$.

Hence, least value is -26 and this occurs when $x = -3$.

Q & A

2

2(a)(i) Express $x^2 - 5x + 8$ in the form $(x + a)^2 + b$, where the values of the constants a and b are to be found.

(ii) Deduce the greatest value of $-x^2 + 5x - 8$. [3]

(b) Solve the simultaneous equations $y = x^2 - x - 7$ and $y = 2x + 3$ algebraically. Write down a geometric interpretation of your results. [5]

(WJEC C1 May 2009 Q4)

Answer

2(a)(i) $x^2 - 5x + 8 = \left(x - \dfrac{5}{2}\right)^2 - \dfrac{25}{4} + 8 = \left(x - \dfrac{5}{2}\right)^2 + \dfrac{7}{4}$

Hence $a = -\dfrac{5}{2}$ and $b = \dfrac{7}{4}$

(ii) $-x^2 + 5x - 8 = -(x^2 - 5x + 8) = -\left[\left(x - \dfrac{5}{2}\right)^2 + \dfrac{7}{4}\right]$

This function is a reflection of the function in part (a) in the x-axis (because of the minus sign).

The function in part (a) has a minimum value at $x = \dfrac{5}{2}$ of $\dfrac{7}{4}$

The reflection in the x-axis will have a maximum value of $-\dfrac{7}{4}$

33

(b)   Equating expressions for y gives:

$x^2 - x - 7 = 2x + 3$

$x^2 - 3x - 10 = 0$

Factorising the quadratic equation gives:

$(x - 5)(x + 2) = 0$

Hence $x = 5$ or -2

The corresponding y-values are found by substituting these two values in $y = 2x + 3$

Hence, when $x = 5$, $y = 13$ and when $x = -2$, $y = -1$.

The equation $y = x^2 - x - 7$ is a curve and the equation $y = 2x + 3$ is a straight line.

The points $(5, 13)$ and $(-2, -1)$ are the points where the curve and line intersect.

3(a)   Given that $k \neq -1$, show that the quadratic equation

$(k + 1)x^2 + 2kx + (k - 1) = 0$

has two distinct real roots.   [4]

(b)   Find the range of values of x satisfying the inequality

$5x^2 + 7x - 6 \leq 0$ [3]

(WJEC C1 May 2009 Q6)

Answer

3(a)   Investigating the discriminant of $(k + 1)x^2 + 2kx + (k - 1) = 0$

Comparing this to $ax^2 + bx + c = 0$

gives $a = k + 1$

$b = 2k$

$c = k - 1$

Discriminant $b^2 - 4ac = (2k)^2 - 4(k + 1)(k - 1)$ | When $b^2 - 4ac > 0$ this implies that there are two real, distinct roots.

$= 4k^2 - 4(k^2 - 1)$

$= 4k^2 - 4k^2 + 4$

$= 4$

Because this value is greater than 0, this means there are two real and distinct roots.

(b) $5x^2 + 7x - 6 \leq 0$

Considering the case where $5x^2 + 7x - 6 = 0$

Factorising gives $(5x - 3)(x + 2) = 0$

Giving the critical values $x = \dfrac{3}{5}$ or -2 (these are the intercepts on the x-axis)

As the curve $y = 5x^2 + 7x - 6$ has a positive coefficient of x^2 the curve will be $\cup$-shaped.

Sketching the curve for $y = 5x^2 + 7x - 6$ gives the following:

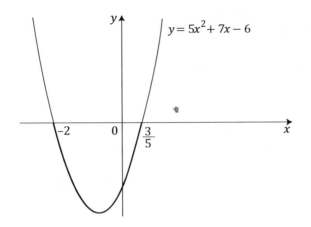

We want the part of the graph which is below or on the x-axis.

Meaning that x lies between -2 and $\dfrac{3}{5}$ inclusive, which can be written mathematically as

$$-2 \leq x \leq \dfrac{3}{5}$$

4

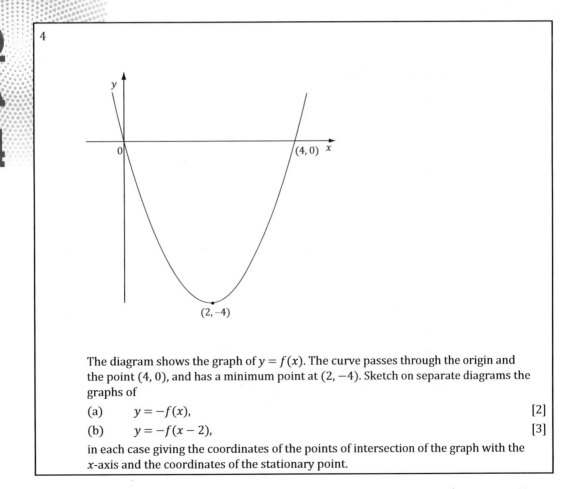

The diagram shows the graph of $y = f(x)$. The curve passes through the origin and the point $(4, 0)$, and has a minimum point at $(2, -4)$. Sketch on separate diagrams the graphs of

(a) $y = -f(x)$,   [2]

(b) $y = -f(x - 2)$,   [3]

in each case giving the coordinates of the points of intersection of the graph with the x-axis and the coordinates of the stationary point.

Answer

4(a) $y = -f(x)$ is a reflection in the x-axis of the graph $y = f(x)$

The points on the x-axis will stay in the same place and the minimum at $(2, -4)$ will be reflected to become a maximum at $(2, 4)$.

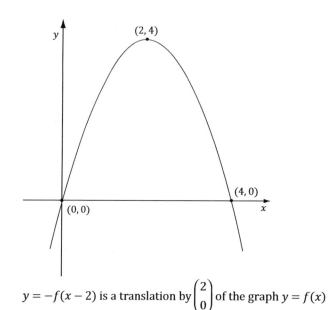

(b) $y = -f(x - 2)$ is a translation by $\begin{pmatrix} 2 \\ 0 \end{pmatrix}$ of the graph $y = f(x)$

The y-coordinates will stay the same but the x-coordinates will be shifted to the right by 2 units.

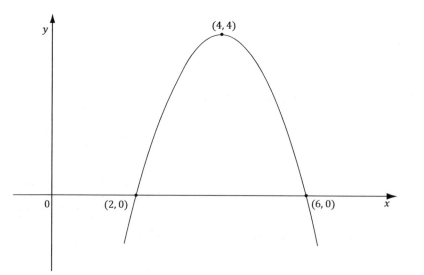

Topic 3 — Coordinate geometry and straight lines

This topic covers the following:

- Finding the gradient, length and mid-point of a straight line
- Finding the equation of a straight line
- Conditions for lines to be parallel or perpendicular to each other

Finding the gradient, equation, length and mid-point of a line joining two points

Finding the gradient of a straight line

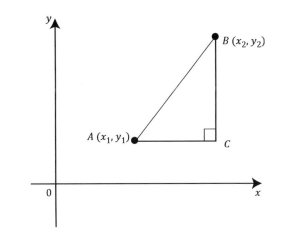

From the above graph length $AC = x_2 - x_1$ and length $BC = y_2 - y_1$

Gradient of line $AB = \dfrac{BC}{AC} = \dfrac{y_2 - y_1}{x_2 - x_1}$.

The gradient of the line joining points (x_1, y_1) and (x_2, y_2) is given by:

$$\text{Gradient} = \frac{y_2 - y_1}{x_2 - x_1}$$

> You need to remember this formula as it will not be given in the formula booklet.

For example the gradient of the straight line AB joining the points $A\ (-3, 2)$ and $B\ (1, 6)$ is

$$\frac{6-2}{1-(-3)} = \frac{4}{4} = 1$$

Finding the length of a straight line joining two points

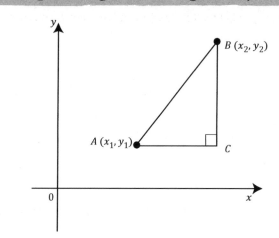

Now $AC = x_2 - x_1$ and length $BC = y_2 - y_1$

By Pythagoras' Theorem $AB^2 = AC^2 + BC^2$

So $AB^2 = (x_2 - x_1)^2 + (y_2 - y_1)^2$

$AB = \sqrt{(x_2 - x_1)^2 + (y_2 - y_1)^2}$

The length of a straight line joining the two points (x_1, y_1) and (x_2, y_2) is given by:

$$\sqrt{(x_2 - x_1)^2 + (y_2 - y_1)^2}$$

Remember this formula.

For example the length of the straight line joining the points A $(-3, -1)$ and B $(1, 2)$ is

$$\sqrt{(1 - (-3))^2 + (2 - (-1))^2} = \sqrt{16 + 9} = \sqrt{25} = 5 \text{ units}$$

Finding the mid-point of a straight line joining two points

The mid-point of a line joining the points (x_1, y_1) and (x_2, y_2) is given by:

$$\left(\frac{x_1 + x_2}{2}, \frac{y_1 + y_2}{2} \right)$$

Remember this formula.

For example the mid-point of the line joining the two points with coordinates $(2, 6)$ and $(8, 4)$ is

$$\left(\frac{2 + 8}{2}, \frac{6 + 4}{2} \right) = (5, 5)$$

Finding the equation of a straight line

To find the equation of a straight line you need to know the gradient of the line (m) and the coordinates of a point (x_1, y_1) that lies on the line.

The equation of a straight line with gradient m and which passes through a point (x_1, y_1) is given by:

$$y - y_1 = m(x - x_1)$$

Remember this formula.

For example, the equation of the straight line with a gradient of 2 and passing through the point (2, 5) is

$$y - 5 = 2(x - 2)$$

$$y - 5 = 2x - 4$$

So $y = 2x + 1$ or $2x - y + 1 = 0$

Notice that the first equation is in the form $y = mx + c$, so you can immediately see that the gradient, $m = 2$ and the intercept on the y-axis, $c = +1$.

The second form gives the equation as $ax + by + c = 0$.

Which form you use depends on whether you are asked for a particular form in the question. You can give the equation of a straight line in either form if a form is not specified in the question.

Example

① Find the equation of the line L, having gradient 3 and passing through the point (2, 3).

Answer

$y - y_1 = m (x - x_1)$ where $m = 3$ and $(x_1, y_1) = (2, 3)$.

Write down the general equation for a straight line and then substitute values into it for m, x_1 and y_1.

$$y - 3 = 3(x - 2)$$

$$y - 3 = 3x - 6$$

This equation is in the form $y = mx + c$, where m is the gradient and c is the intercept on the y-axis.

$$y = 3x - 3$$

Example

② Find the equation of the line in the form $ax + by + c = 0$ that has a gradient of 2 and passes through the point (−1, 0).

Answer

$y - y_1 = m(x - x_1)$ where $m = 2$ and $(x_1, y_1) = (-1, 0)$

$$y - 0 = 2(x - (-1))$$

$$y = 2(x + 1)$$

$$y = 2x + 2$$

Remember to give the equation in the format asked for in the question.

$$2x - y + 2 = 0$$

Example

③ Find the length of the line joining the two points $(-1, -2)$ and $(4, 10)$.

Answer

Using the formula for the distance between two points:

The length of a straight line joining the two points (x_1, y_1) and (x_2, y_2) is given by:

$$\sqrt{(x_2 - x_1)^2 + (y_2 - y_1)^2}$$

> Be careful substituting negative numbers into this formula. It is best to add brackets to emphasise the negative numbers.

Substituting the coordinates $(-1, -2)$ and $(4, 10)$ into this gives

Length $= \sqrt{(4 - (-1))^2 + (10 - (-2))^2} = \sqrt{25 + 144} = \sqrt{169} = 13$ units

Conditions for two straight lines to be parallel or perpendicular to each other

Condition for two straight lines to be parallel to each other

For two lines to be parallel to each other, they must have the same gradient.

For example, the equation of the line that is parallel to the line $y = 3x - 2$ but intersects the y-axis at $y = 2$ is:

$y = 3x + 2$ as $m = 3$ and $c = 2$ (i.e. using the equation $y = mx + c$).

Condition for two straight lines to be perpendicular to each other

When two lines are perpendicular to each other (i.e. they make an angle of 90°), the product of their gradients is -1.

If one line has a gradient m_1 and the other a gradient of m_2 then

$m_1 m_2 = -1$

For example, if a straight line has gradient $-\dfrac{1}{3}$ then the gradient of the line perpendicular to this is given by

$\left(-\dfrac{1}{3}\right) m_2 = -1$, hence gradient $m_2 = 3$

Example

① Find the equation of line L_1 which passes through the point $(1, 2)$ and is parallel to the line L_2 which has equation $2x - y + 1 = 0$.

Answer

① First find the equation of the line L_2 in the form $y = mx + c$

$2x - y + 1 = 0$

> Comparing this equation with $y = mx + c$ gives the gradient, $m = 2$.

So $y = 2x + 1$

Hence the gradient of $L_2 = 2$.

Since lines L_1 and L_2 are parallel, they both have the same gradient of 2.

Finding the equation of line L_1

$y - y_1 = m(x - x_1)$ where $m = 2$ and $(x_1, y_1) = (1, 2)$.

$y - 2 = 2(x - 1)$

$y = 2x$ (or $y - 2x = 0$)

⩕ Grade boost

Always check to see if a form for the equation for the equation of a straight line is given in the question. If the form is not specified, then either equation specified here is acceptable.

Example

② The points A, B and C have coordinates $(1, 1)$, $(3, 3)$ and $(6, 0)$ respectively.

(a) Find the gradients of lines AB and BC.

(b) Prove that lines AB and BC are perpendicular to each other.

Answer

② (a) Gradient of $AB = \dfrac{3 - 1}{3 - 1} = 1$

Gradient of $BC = \dfrac{0 - 3}{6 - 3} = -1$

Both gradients are found using the formula:
Gradient $= \dfrac{y_2 - y_1}{x_2 - x_1}$

(b) Product of the gradients $= (1)\,(-1) = -1$.

As $m_1 m_2 = -1$, AB and BC are perpendicular to each other.

Example

③ The points A, B, C have coordinates $(-11, 10)$, $(-5, 12)$, $(3, 8)$ respectively.
The line L_1 passes through the point A and is **parallel** to BC.
The line L_2 passes through the point C and is **perpendicular** to BC.

(a) Find the gradient of BC. [2]

(b) (i) Show that L_1 has equation: $x + 2y - 9 = 0$.

(ii) Find the equation of L_2. [6]

(c) The lines L_1 and L_2 intersect at the point D.

(i) Show that D has coordinates $(1, 4)$.

(ii) Find the length of BD.

(iii) Find the coordinates of the mid-point of BD. [6]

(WJEC C1 Jan 10 Q1)

Answer

③ (a) Gradient of $BC = \dfrac{8-12}{3-(-5)} = \dfrac{-4}{8} = -\dfrac{1}{2}$

(b) (i) Gradient of line $L_1 = -\dfrac{1}{2}$ as lines L_1 and BC are parallel.

$m = -\dfrac{1}{2}$ and $(x_1, y_1) = (-11, 10)$

Equation of L_1 is given by

$y - y_1 = m(x - x_1)$

$y - 10 = -\dfrac{1}{2}\big(x-(-11)\big)$

$2y - 20 = -x - 11$

$x + 2y - 9 = 0$

(ii) Gradient of line $L_2 = 2$

> Here we use the fact that the product of the gradients of perpendicular lines is –1.

Equation of L_2 is given by $y - y_1 = m(x - x_1)$ where m $= 2$ and $(x_1, y_1) = (3, 8)$

$y - 8 = 2(x - 3)$

$y - 8 = 2x - 6$

$2x - y + 2 = 0$

(c) (i) Solving the equations of lines L_1 and L_2 simultaneously to find the point of intersection:

$x + 2y - 9 = 0$ (1)

$2x - y + 2 = 0$ (2)

> ≫ **Grade boost**
>
> Being able to solve simultaneous equations is assumed at AS level.
>
> You may need to go back to your GCSE work to check you can do them.

Multiplying equation (1) by 2:

$2x + 4y - 18 = 0$

$2x - y + 2 = 0$

Subtracting these two equations gives:

$5y - 20 = 0$

$y = 4$

Substituting $y = 4$ into equation (1) gives:

$x + 8 - 9 = 0$

$x - 1 = 0$

$x = 1$

Checking by substituting the values of x and y into equation (2).

LHS $= 2x - y + 2$

> Always check the values for x and y by substituting them into the equation that you have not used already for the substitution.

$\qquad = 2(1) - 4 + 2 = 0 =$ RHS

Hence D is the point $(1, 4)$

(ii) The length of a straight line joining the two points (x_1, y_1) and (x_2, y_2) is given by:

$$\sqrt{(x_2 - x_1)^2 + (y_2 - y_1)^2}$$

$$\begin{aligned}
\text{Length of } BD &= \sqrt{(-5-1)^2 + (12-4)^2} \\
&= \sqrt{36 + 64} \\
&= \sqrt{100} \\
&= 10
\end{aligned}$$

(iii) The mid-point of a line joining the points (x_1, y_1) and (x_2, y_2) is given by:

$$\left(\frac{x_1 + x_2}{2}, \frac{y_1 + y_2}{2} \right)$$

$$\text{Mid-point of } BD = \left(\frac{-5+1}{2}, \frac{12+4}{2} \right) = (-2, 8)$$

Examination style questions

① A line passes through the points A $(1, -1)$ and B $(3, 4)$.

    (a)   Find the gradient of line AB.   [2]

    (b)   Find the coordinates of C, the mid-point of AB.   [2]

    (c)   The line L is perpendicular to line AB and passes through the point C.   [3]

            Find the equation of line L.

Answer

①   (a) $\text{Gradient} = \dfrac{y_2 - y_1}{x_2 - x_1} = \dfrac{4 - (-1)}{3 - 1} = \dfrac{5}{2}$

    (b)   The mid-point of a line joining the points (x_1, y_1) and (x_2, y_2) is given by:

$$\left(\frac{x_1 + x_2}{2}, \frac{y_1 + y_2}{2} \right)$$

            Hence mid-point of $AB = \left(\dfrac{1+3}{2}, \dfrac{-1+4}{2} \right) = \left(2, \dfrac{3}{2} \right)$

    (c)   The product of the gradients of perpendicular lines is -1. Hence

$$m \left(\frac{5}{2} \right) = -1$$

            Giving $m = -\dfrac{2}{5}$

Equation of straight line L having gradient $-\dfrac{2}{5}$ and passing through the point $\left(2, \dfrac{3}{2}\right)$ is: $y - \dfrac{3}{2} = -\dfrac{2}{5}(x-2)$

> Where there are fractions like this, multiply both sides by the lowest common denominator.

Multiplying through by 10 gives

$10y - 15 = -4(x - 2)$

$10y - 15 = -4x + 8$

$4x + 10y - 23 = 0$

---

② The points A, B, C, D have coordinates $(-4, 4)$, $(-1, 3)$, $(0, 1)$, $(k, 0)$ respectively.

The straight line CD is parallel to the straight line AB.

(a) Find the gradient of AB. [2]

(b) Find the gradient of CD and hence find the value of the constant k. [3]

(c) Line L is perpendicular to CD and passes through point C. Find the equation of line L in the form $ax + by + c = 0$. [2]

Answer

② (a) Gradient of $AB = \dfrac{y_2 - y_1}{x_2 - x_1} = \dfrac{3 - 4}{-1 - (-4)} = -\dfrac{1}{3}$

(b) Gradient of $CD = \dfrac{y_2 - y_1}{x_2 - x_1} = \dfrac{0 - 1}{k - 0} = -\dfrac{1}{k}$

As line CD is parallel to AB the gradients are equal.

Hence, $-\dfrac{1}{3} = -\dfrac{1}{k}$

> The gradients are equated here.

Giving $k = 3$

(c) As line L is perpendicular to CD the product of their gradients is -1.

$\left(-\dfrac{1}{3}\right)m_2 = -1$

Giving gradient of $L = 3$

Equation of line L is:

$y - 1 = 3(x - 0)$

$y = 3x + 1$

Hence $3x - y + 1 = 0$

≫ Grade boost

Leaving the equation in the form $y = mx + c$ (i.e. $y = 3x + 1$) would cost you a mark here as the question asks that the equation be given in the form $ax + by + c = 0$.

Test yourself

Answer the following questions and check your answers before moving on to the next topic.

① The points A, B, C, D have coordinates $(1, 0)$, $(4, 1)$, $(-1, 3)$, $(2, 4)$ respectively.

    (a) Show that lines AB and CD are parallel.

    (b) Find the equation of AB in the form $ax + by + c = 0$.

② The points A and B have coordinates $(-7, 4)$ and $(k, -1)$ respectively.

    (a) If the gradient of AB is $-\dfrac{1}{2}$, find the value of the constant k.

    (b) The line BC is perpendicular to AB. Find the equation of line BC.

③ The points A, B, C have coordinates $(-3, 2)$, $(1, 6)$, $(6, 1)$.

    (a) Show that AB is perpendicular to BC.

    (b) Find the length of AB and the length of BC.

    (c) Find the value of Tan $A\hat{C}B$ in the form $\dfrac{a}{b}$.

(Note: answers to Test yourself are found at the back of the book.)

Q & A 1

1	The points A, B, C, D have coordinates $(-7, 4)$, $(3, -1)$, $(6, 1)$, $(k, -15)$ respectively.	
	(a) Find the gradient of AB.	[2]
	(b) Find the equation of AB and simplify your answer.	[3]
	(c) Find the length of AB.	[2]
	(d) The point E is the mid-point of AB. Find the coordinates of E.	[2]
	(e) Given that CD is perpendicular to AB, find the value of the constant k.	[4]

(WJEC C1 May 08 Q1)

Answer

1(a) Gradient of $AB = \dfrac{y_2 - y_1}{x_2 - x_1} = \dfrac{4-(-1)}{-7-3} = -\dfrac{1}{2}$

> Watch the signs when using this formula. Add brackets to the negative coordinates to emphasise them.

(b) Equation of straight line which passes through $(-7, 4)$ and has gradient $-\frac{1}{2}$ is given by:

$y - y_1 = m(x - x_1)$ where $m = -\frac{1}{2}$ and $(x_1, y_1) = (-7, 4)$

$y - 4 = -\frac{1}{2}\,(x - (-7))$

$2y - 8 = -x - 7$

$x + 2y - 1 = 0$

(c) The length of a straight line joining the two points (x_1, y_1) and (x_2, y_2) is given by:

$$\sqrt{(x_2 - x_1)^2 + (y_2 - y_1)^2}$$

Substituting the coordinates $(-7, 4)$ and $(3, -1)$ into this gives

Length of $AB = \sqrt{(3-(-7))^2 + (-1-4)^2} = \sqrt{100+25} = \sqrt{125} = 5\sqrt{5}$ units

(d) The mid-point of a line joining the points (x_1, y_1) and (x_2, y_2) is given by:

$$\left(\frac{x_1 + x_2}{2}, \frac{y_1 + y_2}{2}\right)$$

Hence mid-point E of $AB = \left(\dfrac{-7+3}{2}, \dfrac{4+(-1)}{2}\right) = \left(-2, \dfrac{3}{2}\right)$

(e) If CD is perpendicular to AB then the product of the gradients equals -1.

$m_1 m_2 = -1$

$\left(-\dfrac{1}{2}\right)m_2 = -1$

Giving gradient of $CD = 2$

Finding the gradients using the coordinates of C (6, 1) and D (k, –15) gives

Gradient of CD $= \dfrac{y_2 - y_1}{x_2 - x_1} = \dfrac{-15 - 1}{k - 6}$

Hence $\dfrac{-15 - 1}{k - 6} = 2$

> The gradient of CD is 2 so an equation can be formed.

$-16 = 2(k - 6)$

$-16 = 2k - 12$

$k = -2$

Topic 4 — Polynomials and the binomial expansion

This topic covers the following:

- Algebraic manipulation of polynomials including expanding brackets and collecting like terms, factorisation and simple algebraic division
- Use of the remainder theorem and factor theorem
- Binomial expansion of $(a + b)^n$ and $(1 + x)^n$ for positive integer values of n.
- Use of $\binom{n}{r}$ and $n!$

Algebraic manipulation of polynomials

In this topic you will cover the manipulation of polynomials such as $27x^3 + 9x^2 - 3x + 7$ and also learn about the binomial expansion of $(1 + x)^n$ for positive integer values of n.

There is a fair amount of algebraic manipulation needed, so you may need to practise some of the skills you learned as part of your GCSE course.

Expanding brackets and collecting like terms

Here are some examples of polynomials. Notice the way they are each ordered in descending powers of x. The degree is the highest power of x in the expression.

$4x - 9$ is a polynomial of degree 1 or a linear expression

$2x^2 + 6x - 1$ is a polynomial of degree 2 or a quadratic expression

$5x^3 + 3x^2 - 2x + 6$ is a polynomial of degree 3 or a cubic expression

You can only add or subtract like terms when simplifying an expression. Like terms are those terms having identical letters and powers. For example $4x^2 y$ cannot be added to $5xy$.

Examples

① $4x^3 + 6x^2y + x^2y + 5xy - xy^2 + xy - 2x^3 = 2x^3 + 7x^2y + 6xy - xy^2$

② $a + ab + ba + b^2 - a - 4b^2 = 2ab - 3b^2$

> ab and ba are the same and can therefore be added.

③ $x(x^2 + 2x - 1) + 2x(x - 3) = x^3 + 2x^2 - x + 2x^2 - 6x$
$$= x^3 + 4x^2 - 7x$$

④ Expand and simplify the brackets in the following expression:

$(x + 2)(x - 3)(x + 4)$

$= (x + 2)(x^2 + 4x - 3x - 12)$

$= (x + 2)(x^2 + x - 12)$

$= x^3 + x^2 - 12x + 2x^2 + 2x - 24 = x^3 + 3x^2 - 10x - 24$

> Here we have multiplied the last pair of brackets first to give a quadratic expression. This is then multiplied by the first bracket to give the final answer. The order in which the brackets are multiplied is unimportant.

Factorisation

Factorisation is the opposite process to expanding brackets. The highest common factor is taken outside the bracket. Each term is divided by the highest common factor and the result is written inside the bracket.

Examples

Factorise the following:

① $x^2y - xy = xy(x - 1)$

② $24x^3y^2z + 6x^2y - 18x^2 = 6x^2(4xy^2z + y - 3)$

③ $15a^2b - 12ab = 3ab(5a - 4)$

The difference of two squares

Both terms must be perfect squares so capable of being easily square-rooted. Note that this only works when there is a minus sign between the two terms.

$$x^2 - y^2 = (x + y)(x - y)$$

$$4x^2 - 9y^2 = (2x + 3y)(2x - 3y)$$

$$16x^2 - 25 = (4x + 5)(4x - 5)$$

> Just square root each term and substitute them in brackets like this with one having a + sign between the terms and the other a − sign.

The following difference of two squares formula should be learnt:

$$a^2 - b^2 = (a + b)(a - b)$$

> ## ⫸ Grade boost
>
> It is always advisable to check your factorisation by multiplying out the brackets. It is easy to make a mistake especially with signs.

Algebraic division

When 25 is divided by 4 the quotient is 6 and the remainder is 1. The number 25 can be written in the following way:

$$25 = 4 \times 6 + 1$$

This can be applied to algebra like this:

Find the quotient and remainder when $x^2 + 5x - 8$ is divided by $x - 2$.

$x^2 + 5x - 8 = (x - 2)(ax + b) + c$ where $ax + b$ is the quotient and c is the remainder.

$$= ax^2 + bx - 2ax - 2b + c$$

$$= ax^2 + (b - 2a)x - 2b + c$$

Comparing this with the original expression and equating the coefficients of x^2 gives

$$a = 1$$

> The coefficients of x, x^2, x^3 etc. are the numbers in front of these terms. The term independent of x is the number without any x's (i.e. the constant term).

Equating coefficients of x gives $5 = b - 2a$ and since $a = 1$, solving gives $b = 7$.

Equating constant terms gives $-2b + c = -8$ and since $b = 7$, $c = 6$.

Hence the quotient (i.e. $ax + b$) is $x + 7$ and the remainder (i.e. c) is 6.

Notice that $x^2 + 5x - 8 = (x - 2)(x + 7) + 6$

The remainder theorem

The remainder theorem states:

If a polynomial $f(x)$ is divided by $(x - a)$ the remainder is $f(a)$.

For example, if $f(x) = x^3 + 2x^2 - x + 1$ is divided by $x - 1$ the remainder will be $f(1)$.

Remainder $= f(1) = 1^3 + 2(1)^2 - 1 + 1 = 3$

Example

① Find the remainder when $x^3 + x^2 + x - 2$ is divided by $x - 1$.

Answer

Let $f(x) = x^3 + x^2 + x - 2$.

If $f(x) = x^3 + x^2 + x - 2$ is divided by $x - 1$, the remainder is $f(1)$.

> This is the remainder theorem.

$f(1) = 1^3 + 1^2 + 1 - 2 = 1$

Hence remainder $= 1$

Example

② Find the remainder when $27x^3 + 9x^2 - 3x + 7$ is divided by $3x - 1$.

> The value of x to be substitute into the function is found by letting $3x - 1 = 0$ and then solving for x giving $x = \dfrac{1}{3}$.

Answer

Let $f(x) = 27x^3 + 9x^2 - 3x + 7$

$$f\left(\frac{1}{3}\right) = 27\left(\frac{1}{3}\right)^3 + 9\left(\frac{1}{3}\right)^2 - 3\left(\frac{1}{3}\right) + 7 = 1 + 1 - 1 + 7 = 8$$

Hence remainder $= 8$.

The factor theorem

A special case of the remainder theorem occurs when there is no remainder, i.e. $f(a) = 0$.

For a polynomial $f(x)$, if $f(a) = 0$ then $(x - a)$ is a factor of $f(x)$.

For example in a polynomial $f(x)$ if $f(5) = 0$, then $(x - 5)$ is a factor of $f(x)$.

If for the same polynomial $f(-2) = 0$, then $(x + 2)$ is also factor of $f(x)$.

Examples

① Prove that $x + 3$ is a factor of the polynomial

$2x^3 + x^2 - 13x + 6$

> If $x + 3$ is a factor, then when $x = -3$ is substitute into $f(x)$ there will be no remainder.

Answer

Let $f(x) = 2x^3 + x^2 - 13x + 6$

If $x + 3$ is a factor then $f(-3)$ should be zero.

$f(-3) = 2(-3)^3 + (-3)^2 - 13(-3) + 6 = -54 + 9 + 39 + 6 = 0$

Hence $x + 3$ is a factor.

② Prove that $x - 2$ is **not** a factor of the function

Answer

$f(x) = 3x^3 - 2x^2 + x + 2$

$f(2) = 3(2)^3 - 2(2)^2 + 2 + 2 = 20$

As $f(2) \neq 0$ then $x - 2$ is not a factor of the function.

> **Grade boost**
>
> Always read the question carefully. It would be easy to miss the word 'not' in this question.

Factorising a polynomial

Suppose a function $f(x)$ is defined by $f(x) = x^3 - 3x^2 - x + 3$

In order to factorise the function it is necessary first to find a factor.

Suppose we think that $(x + 1)$ is a factor. We can see if it is a factor by substituting $x = -1$ into the function. If there is no remainder then $(x + 1)$ is a factor.

$f(-1) = (-1)^3 - 3(-1)^2 - (-1) + 3 = -1 - 3 + 1 + 3 = 0$

Hence $(x + 1)$ is a factor.

> Each term in the first bracket is multiplied by each term in the second bracket.

The function can now be written in the following way:

$f(x) = (x + 1)(ax^2 + bx + c) = x^3 - 3x^2 - x + 3$

Equating coefficients of x^3 gives $a = 1$.

Equating constant terms gives $c = 3$.

Equating coefficients of x gives $c + b = -1$ so $b = -4$.

These values can be substituted in giving:

$f(x) = (x + 1)(x^2 - 4x + 3)$

The second bracket is then factorised giving:

$f(x) = (x + 1)(x - 3)(x - 1)$

> **Grade boost**
>
> Three of the four terms have been equated here. The fourth term could be equated as a check. Here it is the coefficients of x^2. Equating these gives $b + a = -3$. We can substitute the values in $b + a = -4 + 1 = -3$.

A slightly different method of finding the quadratic factor would be to write down the x^2 and the constant term by inspection, e.g. $x^3 - 3x^2 - x + 3 = (x + 1)(x^2 + ax + 3)$ as it is clear that $x \times x^2 = x^3$ and $1 \times 3 = 3$. Then there is only one unknown coefficient to find. Equating coefficients of x^2 or x is sufficient to find this unknown coefficient of x in the quadratic factor – and doing both would act as a useful check.

Example

① The polynomial $f(x)$ is defined by: $\quad f(x) = 2x^3 + 11x^2 + 4x - 5$

   (a)(i)   Evaluate $f(-2)$.

     (ii)   Using your answer to part (i), write down one fact which you can deduce about $f(x)$.   **[2]**

   (b)   Solve the equation $f(x) = 0$ **[6]**

     (WJEC C1 Jan 10 Q8)

Answer

① (a)(i) $f(x) = 2x^3 + 11x^2 + 4x - 5$

        $f(-2) = 2(-2)^3 + 11(-2)^2 + 4(-2) - 5 = 15$

> If $f(-2) = 0$, then $(x + 2)$ is a factor. If there is a remainder, then it is not a factor.

(ii)   Since there is a remainder, this means that $(x + 2)$ is not a factor of $2x^3 + 11x^2 + 4x - 5$

(b)   Using values that are factors of the constant term, (5), we substitute values of x into the function until the function equals zero.

Starting from $f(1), f(-1), f(5)$ etc.

$f(1) = 2(1)^3 + 11(1)^2 + 4(1) - 5 = 12$

$f(-1) = 2(-1)^3 + 11(-1)^2 + 4(-1) - 5 = 0$

Hence $(x + 1)$ is a factor.

> Remember to reverse the sign of the number which gives a zero value, when stating the factor.

As one of the factors is $(x + 1)$ so the original function can be written like this:

$2x^3 + 11x^2 + 4x - 5 = (x + 1)(ax^2 + bx + c)$

Equating the coefficients of x^3 gives, $a = 2$.

Equating the constant terms gives, $c = -5$.

Equating the coefficients of x^2 gives,
$b + a = 11$, so $b = 9$.

Hence, $2x^3 + 11x^2 + 4x - 5 =$
$(x + 1)(2x^2 + 9x - 5)$

Factorising the quadratic part into two factors gives:

$(x + 1)(2x - 1)(x + 5)$

Hence $f(x) = (x + 1)(2x - 1)(x + 5) = 0$

Solutions are $x = -1, \dfrac{1}{2}$ or -5.

> **≫ Grade boost**
>
> You must be able to factorise quadratic expressions quickly and accurately. Practise these using a GCSE text book.

> Substitute each bracket in turn equal to 0 and solve for x to obtain the solutions.

Example

②   (a)   Find the remainder when $x^3 - 17$ is divided by $x - 3$.   [2]

     (b)   Solve the equation $6x^3 - 7x^2 - 14x + 8 = 0$ [6]

        (WJEC C1 Jan 2009 Q7)

Answer

②   (a)   Let $f(x) = x^3 - 17$

        $f(3) = 3^3 - 17 = 10$, hence remainder $= 10$.

     (b)   Let $f(x) = 6x^3 - 7x^2 - 14x + 8$

> You have to use trial and error by substituting values 1, −1, 2, −2, etc., in until you find a value that gives zero when substituted into the function for x. Try values that are factors of the constant term (8).

$f(1) = 6(1)^3 - 7(1)^2 - 14(1) + 8 = -7$

$f(-1) = 6(-1)^3 - 7(-1)^2 - 14(-1) + 8 = 9$

$f(2) = 6(2)^3 - 7(2)^2 - 14(2) + 8 = 0$, hence $(x - 2)$ is a factor is a factor of $f(x)$

$(x - 2)(ax^2 + bx + c) = 6x^3 - 7x^2 - 14x + 8$

Equating coefficients of x^3 gives $a = 6$

Equating constant terms gives $-2c = 8$, so $c = -4$

Equating coefficients of x^2 gives $b - 2a = -7$, so $b = 5$

Hence the equation is factorised to:

$(x - 2)(6x^2 + 5x - 4) = (x - 2)(3x + 4)(2x - 1)$

$(x - 2)(3x + 4)(2x - 1) = 0$

Solving gives $x = 2, -\dfrac{4}{3}$ or $\dfrac{1}{2}$.

Example

③ (a) Given that $x + 2$ is a factor of $12x^3 + kx^2 - 13x - 6$, write down an equation satisfied by k. Hence show that $k = 19$. [2]

(b) Factorise $12x^3 + 19x^2 - 13x - 6$. [3]

(c) Find the remainder when $12x^3 + 19x^2 - 13x - 6$ is divided by $2x - 1$. [2]

(WJEC C1 May 2010 Q8)

Answer

③ (a) If $x + 2$ is a factor, then when $x = -2$ is substitute into the function, the function will equal zero.

Let $f(x) = 12x^3 + kx^2 - 13x - 6$

$f(-2) = 12(-2)^3 + k(-2)^2 - 13(-2) - 6 = 0$

$\qquad -96 + 4k + 26 - 6 = 0$

$\qquad 4k - 76 = 0$

$\qquad k = 19$

(b) Notice that this is the same equation in part (a) with 19 substituted in for k.

Hence we know that $(x + 2)$ is a factor.

So, $(x + 2)(ax^2 + bx + c) = 12x^3 + 19x^2 - 13x - 6$

Equating coefficients of x^3 gives $a = 12$.

Equating coefficients of x^2 gives $b + 2a = 19$

$\qquad\qquad b + 24 = 19$

$\qquad\qquad b = -5$

> With practice the values of a, b and c can be found quickly by inspection.

Equating constant terms gives $2c = -6$ so $c = -3$

Hence $12x^3 + 19x^2 - 13x - 6 = (x + 2)(12x^2 - 5x - 3)$

$\qquad\qquad\qquad\qquad\qquad\quad = (x + 2)(4x - 3)(3x + 1)$

(c) $f(x) = 12x^3 + 19x^2 - 13x - 6$

$f\left(\dfrac{1}{2}\right) = 12\left(\dfrac{1}{2}\right)^3 + 19\left(\dfrac{1}{2}\right)^2 - 13\left(\dfrac{1}{2}\right) - 6$

> $2x - 1 = 0$, so $x = \dfrac{1}{2}$

$\qquad = \dfrac{-25}{4}$

Hence, remainder $= \dfrac{-25}{4}$

Binomial expansion

The binomial expansion is the expansion of an expression of the form $(a + b)^n$ where n is a positive integer.

The formula for the expansion will be given in the formula booklet and is shown here:

$$(a+b)^n = a^n + \binom{n}{1}a^{n-1}b + \binom{n}{1}a^{n-2}b^2 + \ldots + \binom{n}{1}a^{n-r}b^r + \ldots + b^n$$

where $\binom{n}{r} = {}^nC_r = \dfrac{n!}{r!(n-r)!}$

> You do not need to memorise these formulae as they are given in the formula booklet.

$n!$ means n factorial. If $n = 5$ then $5! = 5 \times 4 \times 3 \times 2 \times 1$

Note that $0! = 1$.

Important note: You are not allowed to use calculators in the examination for Core 1.

This means that you will need to be able to substitute numbers into the formula for $\binom{n}{r}$.

≫ Grade boost

It is best not to use calculators to work out ^nC_r when doing questions in class. Get used to substituting the numbers into the formula. Calculators are not allowed in the exam for C1.

Example

To see how this formula is used, we will use an example.

Expand $(a + b)^4$

Answer

First carefully copy down the formula from the formula booklet:

$$(a+b)^n = a^n + \binom{n}{1}a^{n-1}b + \binom{n}{2}a^{n-2}b^2 + \ldots$$

You will also need this formula from the formula booklet.

$$\binom{n}{r} = \frac{n!}{r!(n-r)!}$$

Substituting $n = 4$ into each formula gives:

$$(a+b)^4 = a^4 + \binom{4}{1}a^3b + \binom{4}{2}a^2b^2 + \binom{4}{3}ab^3 + \binom{4}{4}b^4$$

Now substituting the numbers $\binom{4}{1}$ for n and r into $\dfrac{n!}{r!(n-r)!}$ gives $\dfrac{4!}{1!(4-1)!} = \dfrac{4 \times 3 \times 2 \times 1}{3 \times 2 \times 1} = 4$

This is repeated by substituting numbers in for $\binom{4}{2}, \binom{4}{3}$ and $\binom{4}{4}$ giving the numbers 6, 4 and 1 respectively.

Hence: $(a + b)^4 = a^4 + 4a^3b + 6a^2b^2 + 4ab^3 + b^4$

Pascal's triangle

You can also find the coefficients in the expansion of $(a + b)^n$ by using Pascal's triangle.

Suppose you want to expand the expression from the previous example, $(a + b)^4$ using Pascal's triangle.

You would write down Pascal's triangle and look for the line starting 1 and then 4 (because n is 4 here). The line 1, 4, 6, 4, 1 gives the coefficients. This avoids the calculation involving the factorials for each coefficient but you will have to remember how to construct Pascal's triangle.

> Notice that all the rows start and end with a 1. Notice also that the other numbers are found by adding the pairs of numbers immediately above. For example if we have 1 3 in the line above then the number to be entered between these numbers on the next line is a 4.

```
              1
           1     1
        1     2     1
     1     3     3     1
  1     4     6     4     1
1     5    10    10     5     1
```

⋀ Grade boost

If you intend to use Pascal's triangle, you must remember how to construct it and also how to decide which line should be used. You will not be given Pascal's triangle in the formula booklet.

Example

Use the binomial expansion to expand $(2 + 3x)^3$.

Answer

First obtain the formula for the binomial expansion from the formula booklet.

$$(a + b)^n = a^n + \binom{n}{1}a^{n-1}b + \binom{n}{2}a^{n-2}b^2 + \ldots$$

Here $a = 2$, $b = 3x$ and $n = 3$.

Substituting these values into the formula gives:

$$(2 + 3x)^3 = 2^3 + \binom{3}{1}2^2(3x) + \binom{3}{2}2^1(3x)^2 + \binom{3}{3}2^0(3x)^3$$

As $n = 3$ here we look for the line in Pascal's triangle which starts at 1 and then 3, etc.

You can see that the numbers in this line are: 1 3 3 1

These are the values of $\binom{3}{0}, \binom{3}{1}, \binom{3}{2}$ and $\binom{3}{3}$. So for example $\binom{3}{1} = 3$ and $\binom{3}{3} = 1$.

Hence we can write the expansion like this:

$$(2 + 3x)^3 = (1)2^3 + (3)2^2(3x) + (3)2^1(3x)^2 + (1)2^0(3x)^3$$

> Remember that $2^0 = 1$

Hence $(2 + 3x)^3 = 8 + 36x + 54x^2 + 27x^3$

The binomial expansion where $a = 1$

When the first term in the bracket (i.e. a) is 1, the binomial expansion becomes:

$$(1 + x)^n = 1 + nx + \frac{n(n-1)}{2!}x^2 + \frac{n(n-1)(n-2)}{3!}x^3 + \ldots$$

Again this formula is given in the formula booklet so you don't need to memorise it.

Example

① (a) Write down the expansion of $(1 + x)^6$ in ascending powers of x up to and including the term in x^3. [2]

(b) By substituting an appropriate value for x in your expansion in (a), find an approximate value for 0.99^6. Show all your working and give your answer correct to four decimal places. [3]

(WJEC C1 May 2010 Q4)

Answer

① (a) The formula for the expansion of $(1 + x)^n$ is obtained from the formula booklet.

$$(1+x)^n = 1 + nx + \frac{n(n-1)}{2!}x^2 + \frac{n(n-1)(n-2)}{3!}x^3 + \dots$$

Substituting $n = 6$ into this formula gives:

$$(1+x)^6 = 1 + 6x + \frac{6(5)}{2!}x^2 + \frac{6(5)(4)}{3!}x^3 + \dots$$

Note that using the first three terms only provides an approximate value.

Hence, $(1+x)^6 \approx 1 + 6x + \frac{6(5)x^2}{2!} + \frac{6(5)(4)x^3}{3!}$

$$\approx 1 + 6x + 15x^2 + 20x^3$$

(b) $1 - 0.01 = 0.99$
So, $0.99^6 = (1 - 0.01)^6$

Substituting $x = -0.01$ into the expansion of $(1 + x)^6$ gives

$(1 - 0.01)^6 \approx 1 + 6(-0.01) + 15(-0.01)^2 + 20(-0.01)^3$

$$\approx 0.94148$$

$$\approx 0.9415 \text{ (4 decimal places)}$$

Grade boost

When obtaining a numerical answer, always check to see if the question asks for the answer to be given to a certain number of decimal places or significant figures. Marks can be needlessly lost by not doing this.

$a = x$
$b = \frac{2}{x}$

Example

② (a) Expand $\left(x + \dfrac{2}{x}\right)^4$, simplifying each term of the expansion. [4]

(b) The coefficient of x^2 in the expansion of $(1 + x)^n$ is 55. Given that n is a positive integer, find the value of n. [3]

(WJEC C1 May 2009 Q7)

Answer

② (a) Obtaining the formula and following the pattern in the terms gives:

$$(a+b)^n = a^n + \binom{n}{1}a^{n-1}b + \binom{n}{2}a^{n-2}b^2 + \binom{n}{3}a^{n-3}b^3 + \dots$$

$$(a+b)^4 = a^4 + \binom{4}{1}a^3b + \binom{4}{2}a^2b^2 + \binom{4}{3}ab^3 + \binom{4}{4}b^4$$

$$(a+b)^4 = a^4 + 4a^3b + 6a^2b^2 + 4ab^3 + b^4$$

Substituting $a = x$ and $b = \dfrac{2}{x}$ into the equation,

gives:

$$\left(x + \frac{2}{x}\right)^4 = x^4 + 4x^3\left(\frac{2}{x}\right) + 6x^2\left(\frac{2}{x}\right)^2 + 4x\left(\frac{2}{x}\right)^3 + \left(\frac{2}{x}\right)^4$$

$$= x^4 + 8x^2 + 24 + \frac{32}{x^2} + \frac{16}{x^4}$$

> **Grade boost**
>
> You can find these numbers using Pascal's triangle but you will need to know how to construct and use it, as it is not given in the formula booklet.

> With practice the values of the coefficients will become known or can be calculated quickly.

(b) In the expansion of $(1 + x)^n$ the coefficient of x^2 is $\dfrac{n(n-1)}{2}$

Hence $\dfrac{n(n-1)}{2} = 55$

$n^2 - n = 110$

$n^2 - n - 110 = 0$

> Notice that this is a quadratic equation so it needs to be rearranged to equal zero so it can be factorised and solved.

Factorising this quadratic equation gives $(n-11)(n+10) = 0$

Solving gives $n = 11$ or -10

The question says n is a positive integer, so $n = 11$.

Examination style questions

① (a) Given that $x - 2$ is a factor of $x^3 - 6x^2 + ax - 6$, show that $a = 11$. [2]

(b) Solve the equation $x^3 - 6x^2 + 11x - 6 = 0$ [4]

(c) Calculate the remainder when $x^3 - 6x^2 + 11x - 6$ is divided by $x + 1$. [2]

Answer

① (a) Let $f(x) = x^3 - 6x^2 + ax - 6$

$f(2) = 2^3 - 6(2)^2 + 2a - 6 = 2a - 22$

As $x - 2$ is a factor, $f(2) = 0$

Hence $2a - 22 = 0$ so $a = 11$.

(b) $f(x) = x^3 - 6x^2 + 11x - 6 = (x-2)(ax^2 + bx + c)$

Equating coefficients of x^3 gives $a = 1$.

Equating constant terms gives $-2c = -6$, giving $c = 3$.

> Always look back at the previous part to see if it is relevant. Here it is because the polynomial is the same and we know that $x - 2$ is a factor.

Equating coefficients of x^2 gives $b - 2a = -6$, giving $b = -4$.

Hence, $f(x) = x^3 - 6x^2 + 11x - 6 = (x - 2)(x^2 - 4x + 3)$

Factorising the second bracket gives:

$f(x) = (x - 2)(x - 3)(x - 1)$

Now $(x - 2)(x - 3)(x - 1) = 0$

So $x = 2, x = 3, x = 1$

(c) $f(-1) = (-1)^3 - 6(-1)^2 + 11(-1) - 6 = -1 - 6 - 11 - 6 = -24$

Remainder $= -24$.

---

② Write down and simplify the first four terms in the binomial expansion of $\left(1 + \dfrac{x}{2}\right)^6$.   [4]

Answer

$(1 + x)^n = 1 + nx + \dfrac{n(n-1)}{2!}x^2 + \dfrac{n(n-1)(n-2)}{3!}x^3$

> Substitute n as 6 and x as $\left(\dfrac{x}{2}\right)$ into the formula.

$\left(1 + \dfrac{x}{2}\right)^6 = 1 + 6\left(\dfrac{x}{2}\right) + \dfrac{(6)(5)}{2 \times 1}\left(\dfrac{x}{2}\right)^2 + \dfrac{(6)(5)(4)}{3 \times 2 \times 1}\left(\dfrac{x}{2}\right)^3$

$= 1 + 3x + \dfrac{15}{4}x^2 + \dfrac{5}{2}x^3$

$x^2 = 4x$

---

③ In the binomial expansion of $(a + 2x)^5$, the coefficient of the term in x^2 is four times the coefficient of the term in x. Find the value of the constant a.   [3]

Answer

$(a + b)^n = a^n + \binom{n}{1}a^{n-1}b + \binom{n}{2}a^{n-2}b^2 + \ldots$

Here $a = a$, $b = 2x$ and $n = 5$.

Substituting these values into the formula gives:

$(a + 2x)^5 = a^5 + \binom{5}{1}a^4(2x) + \binom{5}{2}a^3(2x)^2 + \ldots$

Now $\binom{5}{1} = \dfrac{5!}{1!(5-1)!} = \dfrac{5!}{1!4!} = 5$ and

$\binom{5}{2} = \dfrac{5!}{2!(5-2)!} = \dfrac{5!}{2!3!} = 10$

Hence:

$(a + 2x)^5 = a^5 + (5)a^4(2x) + (10)a^3(2x)^2 + \ldots$

$= a^5 + 10a^4x + 40a^3x^2 + \ldots$

Grade boost

You could have used Pascal's triangle here but this is not in the formula booklet so you would need to know how to construct it.

> The formula $\binom{n}{r} = \dfrac{n!}{r!(n-r)!}$ has been used here. This formula is obtained from the formula booklet.

$n!$

$r!(n-r)!$

Coefficient of x^2 is four times the coefficient of x, so

$40a^3 = 4 \times 10a^4$

$40a^3 = 40a^4$

Dividing both sides by $40a^3$ gives $a = 1$. $(a \neq 0)$

---

④ Expand $(a+b)^4$. Hence expand $\left(2x + \dfrac{1}{2x}\right)^4$, simplifying each term of the expansion. [4]

Answer

④ $(a+b)^n = a^n + \binom{n}{1}a^{n-1}b + \binom{n}{2}a^{n-2}b^2 + \binom{n}{3}a^{n-3}b^3 + \ldots$

$(a+b)^4 = a^4 + \binom{4}{1}a^3b + \binom{4}{2}a^2b^2 + \binom{4}{3}ab^3 + \binom{4}{4}b^4$

Finding $\binom{4}{1}, \binom{4}{2}, \binom{4}{3}, \binom{4}{4}$ by using the formula or by using Pascal's triangle and substituting them in to the above formula gives:

$(a+b)^4 = a^4 + 4a^3b + 6a^2b^2 + 4ab^3 + b^4$

$\left(2x - \dfrac{1}{2x}\right)^4 = (2x)^4 + 4(2x)^3\left(\dfrac{1}{2x}\right) + 6(2x)^2\left(\dfrac{1}{2x}\right)^2 + 4(2x)\left(\dfrac{1}{2x}\right)^3 + \left(\dfrac{1}{2x}\right)^4$

$= 16x^4 + 16x^2 + 6 + \dfrac{1}{x^2} + \dfrac{1}{16x^4}$

Test yourself

Answer the following questions and check your answers before moving on to the next topic.

① Calculate the remainder when $4x^3 + 3x^2 - 3x + 1$ is divided by $x + 1$.

② (a) Given that $x + 2$ is a factor of $x^3 + 6x^2 + ax + 6$, show that $a = 11$.

    (b) Solve the equation $x^3 + 6x^2 + 11x + 6 = 0$.

③ The polynomial $f(x)$ is defined by: $f(x) = x^3 - x^2 - 4x + 4$

    (a) (i) Evaluate $f(-2)$.

        (ii) Using your answer to part (i). Write down one fact which you can deduce about $f(x)$.

    (b) Solve the equation $f(x) = 0$.

④ In the binomial expansion of $(2 + 3x)^5$, find the coefficient of the term in x^2.

⑤ Write down and simplify the first four terms in the binomial expansion of $(1 + 3x)^6$

(Note: answers to Test yourself are found at the back of the book.)

Q&A 1

| 1 | Use the binomial theorem to expand $(3 + 2x)^3$, simplifying each term of your expansion. | [3] |

Grade boost

You have to use the binomial theorem here as it is specified in the question. If you found the answer my multiplying out the brackets you would not gain any marks.

Answer

1   Obtaining the formula gives:

$$(a+b)^n = a^n + na^{n-1}b + \frac{n(n-1)}{2!}a^{n-2}b^2 + \frac{n(n-1)(n-2)}{3!}a^{n-3}b^3$$

Here $n = 3$, $a = 3$ and $b = 2x$.

$$(3+2x)^3 = 3^3 + 3(3)^2(2x) + \frac{(3)(2)}{2!}3^1(2x)^2 + \frac{(3)(2)(1)}{3!}3^0(2x)^3$$

$$= 27 + 54x + 36x^2 + 8x^3$$

Q&A 2

2	The polynomial $4x^3 + px^2 - 11x + q$ has $x - 2$ as a factor. When the polynomial is divided by $x + 1$, the remainder is 9.	
(a)	Show that $p = -4$ and $q = 6$.	[6]
(b)	Factorise $4x^3 - 4x^2 - 11x + 6$.	[3]

(WJEC C1 May 2008 Q7)

Answer

2   (a)   As $x - 2$ is a factor, when $x = 2$ is substituted into the function, the result will be zero.

Let $f(x) = 4x^3 + px^2 - 11x + q$

$f(2) = 4(2)^3 + p(2)^2 - 11(2) + q = 10 + 4p + q$

This equals zero so:

$10 + 4p + q = 0$ (1)

Also, when the polynomial is divided by $(x + 1)$ it gives a remainder of 9.

$f(-1) = 4(-1)^3 + p(-1)^2 - 11(-1) + q$

$= -4 + p + 11 + q$

$= 7 + p + q$

This remainder equals 9 so:

$9 = 7 + p + q$

$2 = p + q$ (2)

Solving equations (1) and (2) simultaneously:

From equation (2) $q = 2 - p$

Substituting this into equation (1) gives

$10 + 4p + 2 - p = 0$

$12 + 3p = 0$

$p = -4$

Substituting $p = -4$ into equation (2) gives

$2 = -4 + q$

$q = 6$

(b)   You know that $x - 2$ is a factor of
$4x^3 - 4x^2 - 11x + 6$

So $(x - 2)(ax^2 + bx + c) = 4x^3 - 4x^2 - 11x + 6$

Equating coefficients of x^3 gives $a = 4$.

Equating constant terms gives $-2c = 6$ so $c = -3$.

Equating coefficients of x^2 gives $b - 2a = -4$

$$b - 8 = -4$$

$$b = 4$$

Substituting these values gives:

$(x - 2)(4x^2 + 4x - 3)$

Factorising the second bracket gives:

$(x - 2)(2x + 3)(2x - 1)$

> **⟰ Grade boost**
>
> Check you can solve simultaneous equations. You may need to revise your GCSE work.

> Always look back at previous parts to see if they are relevant. Here the previous part is relevant.

> The second bracket contains a quadratic which can be factorised.

Topic 5	Differentiation

This topic covers the following:

- Differentiation from first principles
- Differentiation of x^n and related sums and differences
- Stationary points
- The second derivative
- Increasing and decreasing functions
- Simple optimisation problems
- Gradients of tangents and normals, and their equations
- Simple curve sketching

What is differentiation?

Unlike a straight line, which has a fixed gradient, a curve has a variable gradient depending on the point on the curve where the gradient is taken. The gradient at a point of the curve is the gradient of the tangent to the curve at that point. A tangent is a straight line that touches the curve at a point $P(x, y)$.

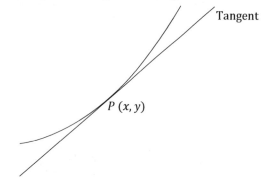

Differentiation is the process of finding a general expression for the gradient of a curve at any point. This general expression for the gradient is known as the derivative, and can be expressed in two ways: $\dfrac{dy}{dx}$ or $f'(x)$

Differentiation from first principles

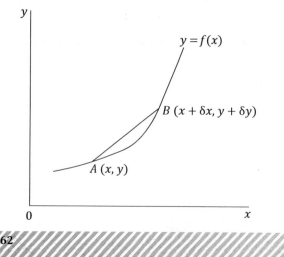

The line joining points A and B is called a chord. Notice that there is a small horizontal distance δx and a small vertical distance δy between points A and B. As A and B move closer together the gradient of the chord AB becomes nearer the true gradient of the tangent to the curve at point A. As $\delta x \to 0$ (i.e. as δx approaches zero) the chord will tend to become the tangent to the curve at point A and the gradient of the curve at point A will be the limit of the gradient of the chord.

This can be expressed in the following way:

$$\frac{dy}{dx} = \lim_{\delta x \to 0} \frac{\delta y}{\delta x} = \lim_{\delta x \to 0} \left(\frac{f(x + \delta x) - f(x)}{\delta x} \right)$$

You have to be able to differentiate from first principles a polynomial of degree less than three (i.e. up to and including terms in x^2).

Suppose we want to find $\dfrac{dy}{dx}$ for $y = 4x^2 - 2x + 1$

Increasing x by a small amount δx will result in y increasing by a small amount δy.

Substituting $x + \delta x$ and $y + \delta x$ into the equation we have:

$y + \delta y = 4(x + \delta x)^2 - 2(x + \delta x) + 1$

$y + \delta y = 4(x^2 + 2x\delta x + (\delta x)^2) - 2x - 2\delta x + 1$

$y + \delta y = 4x^2 + 8x\delta x + 4(\delta x)^2 - 2x - 2\delta x + 1$

But $y = 4x^2 - 2x + 1$

Subtracting these equations gives

$\delta y = 8x\delta x + 4(\delta x)^2 - 2\delta x$

Dividing both sides by δx

$\dfrac{\delta y}{\delta x} = 8x + 4\delta x - 2$

Letting $\delta x \to 0$

$\dfrac{dy}{dx} = \lim_{\delta x \to 0} \dfrac{\delta y}{\delta x} = 8x - 2$

≫ Grade boost

You must include this step about the limits. Leaving this step out will cost you a mark.

Differentiation of x^n and related sums and differences

Before differentiating an expression, it needs to be written in index form. You may need to look back at Topic 1 to revise indices.

To differentiate an expression: multiply by the index and then reduce the index by one.

If $y = kx^n$ then the derivative $\dfrac{dy}{dx} = nkx^{n-1}$

Examples

① If $y = 6x^3 + \dfrac{1}{2}x^2 - 5x + 4$, find $\dfrac{dy}{dx}$.

Differentiating gives

$$\frac{dy}{dx} = (3)6x^2 + (2)\frac{1}{2}x - 5$$

$$\frac{dy}{dx} = 18x^2 + x - 5$$

> ## Grade boost
>
> This step shows the working. You should show your working because if you make an arithmetic error, then you may still get marks for your method.

② Find the gradient of the curve $y = 3x^2 - x + 2$ at the point $P(2, 12)$.

Differentiating the equation of the curve gives

$$\frac{dy}{dx} = (2)3x^1 - 1 = 6x - 1$$

> When differentiating a term in x you obtain the coefficient of x (e.g. $5x$ differentiated becomes 5). A number on its own when differentiated becomes zero.

At P, $x = 2$ so gradient $\dfrac{dy}{dx} = 6(2) - 1 = 11$

> The x value of the point on the curve where the gradient is to be found is substitute into the expression for $\dfrac{dy}{dx}$

③ Given that $y = \sqrt{x} + \dfrac{4}{x^3} + 4$, find the value of $\dfrac{dy}{dx}$ when $x = 1$.

Writing the equation in index form gives

$$y = x^{\frac{1}{2}} + 4x^{-3} + 4$$

Differentiating gives

> Be careful here as a common mistake is to convert to index form and then forget to differentiate the result.

$$\frac{dy}{dx} = \frac{1}{2}x^{-\frac{1}{2}} + (-3)4x^{-4} = \frac{1}{2}x^{-\frac{1}{2}} - 12x^{-4}$$

> When differentiating a term with a negative index, the index of the derivative will still be one less, e.g. if $y = x^{-3}$ then $\dfrac{dy}{dx} = -3x^{-4}$.

Writing this in a form in which numbers are easily entered gives

$$\frac{dy}{dx} = \frac{1}{2\sqrt{x}} - \frac{12}{x^4}$$

Substituting $x = 1$ gives

$$\frac{dy}{dx} = \frac{1}{2\sqrt{1}} - \frac{12}{1^4} = \frac{1}{2} - 12 = -11.5$$

④ If $f(x) = \dfrac{3}{4}x^{\frac{1}{3}} + \dfrac{12}{x^2}$, find the value of $f'(x)$ when $x = 8$.

> The derivative of $f(x)$ is written as $f'(x)$.

Writing the whole function in index form gives

$$f(x) = \frac{3}{4}x^{\frac{1}{3}} + 12x^{-2}$$

> You need to be confident in reducing fractional indices by 1, (e.g. $\dfrac{1}{2} - 1 = -\dfrac{1}{2}$, $-\dfrac{1}{2} - 1 = -\dfrac{3}{2}$, etc)

Differentiating gives

$$f'(x) = \left(\frac{1}{3}\right)\frac{3}{4}x^{-\frac{2}{3}} + (-2)12x^{-3} = \frac{1}{4}x^{-\frac{2}{3}} - 24x^{-3}$$

Writing this in non-index form, gives:

$$f'(x) = \frac{1}{4\sqrt[3]{x^2}} - \frac{24}{x^3}$$

Hence, $f'(8) = \frac{1}{4\sqrt[3]{8^2}} - \frac{24}{8^3} = \frac{1}{16} - \frac{3}{64} = \frac{4}{64} - \frac{3}{64} = \frac{1}{64}$

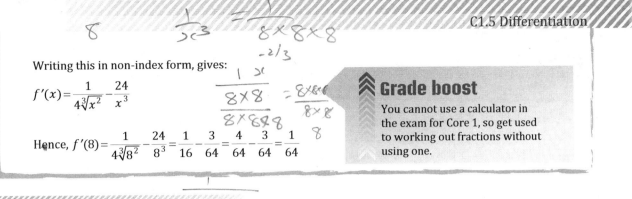

Stationary points

A stationary point is a point on a curve where the gradient is zero. A tangent to the curve at a stationary point will have zero gradient and therefore be parallel to the x-axis.

To find the stationary points on a curve you first differentiate the equation of the curve and then substitute the derivative equal to zero. The resulting equation is solved to find the x-coordinate or coordinates of the stationary points.

Maximum and minimum points

Look carefully at the graph drawn here and notice the way the sign of the gradient changes either side of a stationary point for a maximum and minimum.

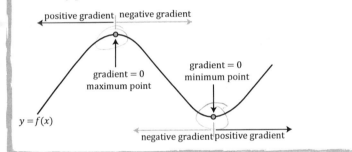

Point of inflection

A point of inflection is a stationary point on a curve (i.e. where the gradient is zero) but where the gradient does not change either side of the stationary point.

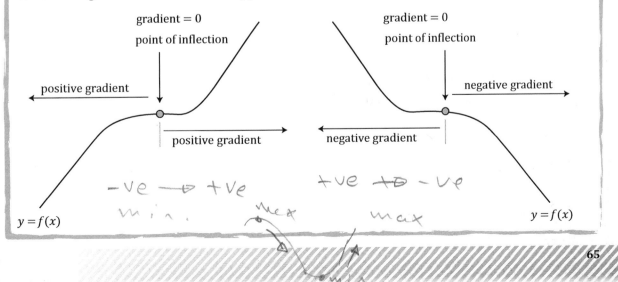

The second order derivative

In order to find the second derivative (i.e. $\dfrac{d^2y}{dx^2}$ or $f''(x)$) you take the first derivative (i.e. $\dfrac{dy}{dx}$ or $f'(x)$) and differentiate it again.

The second derivative gives the following information about the stationary points:

If $\dfrac{d^2y}{dx^2}$ or $f''(x) < 0$ the point is a maximum point.

If $\dfrac{d^2y}{dx^2}$ or $f''(x) > 0$ the point is a minimum point.

If $\dfrac{d^2y}{dx^2}$ or $f''(x) = 0$ this gives no further information about the nature of the point and further investigation is necessary.

Example

① The curve C has equation: $y = -2x^3 + 3x^2 + 12x - 5$

Find the coordinates and nature of each of the stationary points of C. [6]

Answer

① $y = -2x^3 + 3x^2 + 12x - 5$

$$\frac{dy}{dx} = -6x^2 + 6x + 12 = -6(x^2 - x - 2) = -6(x - 2)(x + 1)$$

> When factorising, care should be taken when extracting a negative number as shown here. A common error is simply to divide through by a common factor before equating to zero, which can result in a change of sign – which in turn could lead to a mis-identification of the nature of any stationary points.

At the stationary points, $\dfrac{dy}{dx} = 0$, so

$-6(x - 2)(x + 1) = 0$

Solving gives $x = 2$ or -1

To find the corresponding y values, each of these values is substitute into the equation for the curve.

When $x = 2$, $y = -2(2)^3 + 3(2)^2 + 12(2) - 5 = 15$

When $x = -1$, $y = -2(-1)^3 + 3(-1)^2 + 12(-1) - 5 = -12$

Stationary points are $(2, 15)$ and $(-1, -12)$

To find the nature of the stationary points, $\dfrac{dy}{dx}$ is differentiated again.

$$\frac{d^2y}{dx^2} = -12x + 6$$

Each x value is entered in turn to determine whether the second derivative is positive or negative. If negative, the point is a maximum, and if positive, the point is a minimum.

Hence when $x = 2$, $\dfrac{d^2y}{dx^2} = -12(2) + 6 = -18 < 0$ showing there is a maximum point when $x = 2$.

When $x = -1$, $\dfrac{d^2y}{dx^2} = -12(-1) + 6 = 18 > 0$ showing there is a minimum point when $x = -1$.

Hence (2, 15) is a maximum point and (−1, −12) is a minimum point.

Increasing and decreasing functions

Curves have changing gradients. Depending on the x-coordinate of a point on the curve the gradient, as given by $\dfrac{dy}{dx}$ or $f'(x)$, can have a positive, negative or zero value.

You may have to show that a particular function is increasing or decreasing at a given point. To do this, you find the gradient by differentiating the equation of the curve and then substituting in the x-coordinate of the given point to see whether the gradient is positive or negative.

If the gradient is positive, then the curve at that point is an increasing function.

If the gradient is negative, then the curve at that point is a decreasing function.

Example

A curve C has the equation $y = x^3 - 6x^2 + 2x - 1$

Determine whether y is an increasing or decreasing function at $x = 2$.

Answer

Differentiating gives $\dfrac{dy}{dx} = 3x^2 - 12x + 2$

When $x = 2$, $\dfrac{dy}{dx} = 3(2)^2 - 12(2) + 2 = 12 - 24 + 2 = -10$

The gradient at $x = 2$ is negative, showing that y is a decreasing function at this point.

Simple optimisation problems

Suppose you have the following problem:

You are given a rectangular sheet of metal having dimensions 16 cm by 10 cm.

A square of metal is to be cut out of each corner as shown in the following diagram.

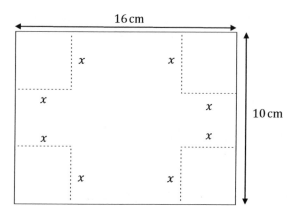

After removing the corners and folding up the flaps an open top box is formed like that shown here:

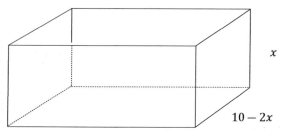

Find the value of x that will make the volume of the box a maximum, and find this maximum volume.

Volume of the box, $V = (16 - 2x)(10 - 2x)(x) = (160 - 52x + 4x^2)(x)$

$$= 160x - 52x^2 + 4x^3$$

Differentiating V with respect to x gives the following:

$$\frac{dV}{dx} = 160 - 104x + 12x^2 = 4(3x^2 - 26x + 40) = 4(3x - 20)(x - 2)$$

For stationary points $\dfrac{dV}{dx} = 0$

$4(3x - 20)(x - 2) = 0$

Giving $x = \dfrac{20}{3}$ or 2

$x = \dfrac{20}{3} = 6\dfrac{2}{3}$ is an impossible answer as the box is only 10 cm wide so it is not possible to cut two squares of side $6\dfrac{2}{3}$ cm.

Hence $x = 2$ cm

We can check that $x = 2$ is a maximum value by finding the second derivative and then substituting 2 in for x to check that the second derivative is negative.

$$\frac{d^2V}{dx^2} = -104 + 24x$$

When $x = 2$, $\frac{d^2V}{dx^2} = -104 + 24(2) = -56$. This is a negative value so the maximum value of V occurs when $x = 2$.

Substituting $x = 2$ into the equation for the volume gives:

$$V = 160(2) - 52(2)^2 + 4(2)^3 = 144$$

Hence, the maximum volume of the box $= 144$ cm^3

Determining whether a stationary point is a point of inflection

The gradient (i.e. $\frac{dy}{dx}$ or $f'(x)$) at a stationary point is zero.
At a point of inflection there is no change in the sign of the gradient either side of the stationary point.

Example

Curve C has the equation $y = x^3 - 6x^2 + 12x - 5$.

Find the coordinates of the stationary point on curve C and show that this point is a point of inflection.

Answer

Differentiating gives $\frac{dy}{dx} = 3x^2 - 12x + 12 = 3(x^2 - 4x + 4) = 3(x - 2)^2$

At the stationary point, $\frac{dy}{dx} = 0$, so $3(x - 2)^2 = 0$

Solving gives a turning point when $x = 2$

To find the y-coordinate of the stationary point, $x = 2$ is substituted into the equation for the curve.

When $x = 2$, $y = 2^3 - 6(2)^2 + 12(2) - 5 = 3$.

Hence the stationary point is at $(2, 3)$.

To show that this is a point of inflection we find the gradient either side of the stationary point and show that the gradient does not change its sign.

$\frac{dy}{dx} = 3(x - 2)^2 \geq 0$ for all values of x, since any expression squared cannot be negative.

The gradient does not change sign so the stationary point at $x = 2$ is a point of inflection.

Gradients of tangents and normals, and their equations

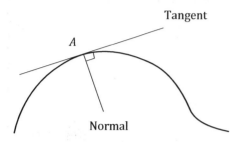

The tangent to a curve and the normal to the curve at the same point are perpendicular to each other.

The gradient of the tangent at point A is the same as the gradient of the curve at point A.

If two lines are perpendicular, the product of their gradients is –1.

> You may need to look back at Topic 3 on coordinate geometry and straight lines before looking at the rest of this section.

To find the equation of the tangent to a curve at a point $P(x, y)$

1 Differentiate the equation of the curve to find the gradient $\dfrac{dy}{dx}$.

2 Substitute the x-coordinate of P into $\dfrac{dy}{dx}$ to obtain the gradient of the tangent at P, m.

3 Using the formula for the equation of a straight line: $y - y_1 = m(x - x_1)$, substitute the gradient m, and the coordinates of point P for x_1 and y_1 into the above formula. Rearrange the equation if necessary so that equation is in the format asked for in the question.

To find the equation of a normal to a curve at a point $P(x, y)$

1 Differentiate the equation of the curve to find the gradient $\dfrac{dy}{dx}$.

2 Substitute the x-coordinate of P into $\dfrac{dy}{dx}$ to obtain the gradient of the tangent at P, m_1.

3 Find the gradient of the normal using $m_1 m_2 = -1$, i.e. $m_2 = -\dfrac{1}{m_1}$

4 Using the formula for the equation of a straight line.
$y - y_1 = m(x - x_1)$, substitute the gradient m_2, and the coordinates of point P for x_1 and y_1 into the above formula. Rearrange the equation if necessary so that equation is in the format asked for in the question.

Example

① The curve C has equation $y = \dfrac{6}{x^2} + \dfrac{7x}{4} - 2$. The point P has coordinates $(2, 3)$ and lies on C.

Find the equation of the **normal** to C at P. [6]

Answer

① $y = \dfrac{6}{x^2} + \dfrac{7x}{4} - 2$

Substituting this equation in index form gives

$y = 6x^{-2} + \dfrac{7x}{4} - 2$

> Reducing the index –2, by 1 gives –3

$\dfrac{dy}{dx} = -12x^{-3} + \dfrac{7}{4}$

$\dfrac{dy}{dx} = -\dfrac{12}{x^3} + \dfrac{7}{4}$

When $x = 2$

$\dfrac{dy}{dx} = -\dfrac{12}{8} + \dfrac{7}{4}$

$= \dfrac{1}{4}$

> This is the numerical value for the gradient of the tangent. This is then used to determine the gradient of the normal which is at right-angles to the tangent.

To find the gradient of the normal we use $m_1 m_2 = -1$

So, $\left(\dfrac{1}{4}\right) m_2 = -1$ $(m_2$ is the gradient of the normal)

Giving gradient of the normal, $m_2 = -4$

Equation of a straight line having gradient m and passing through the point (x_1, y_1) is given by:

$y - y_1 = m(x - x_1)$ where $m = -4$ and $(x_1, y_1) = (2, 3)$, so

$y - 3 = -4(x - 2)$

$y = -4x + 11$

Example

② The curve C has equation $y = \dfrac{1}{2}x^3 - 6x + 3$

Find the coordinates and the nature of each of the stationary points of C. [6]

(WJEC C1 May 2010 Q10)

Answer

② $y = \dfrac{1}{2}x^3 - 6x + 3$

Differentiating gives $\dfrac{dy}{dx} = \dfrac{3}{2}x^2 - 6$

At the stationary point, $\dfrac{dy}{dx} = 0$

Hence $\dfrac{3}{2}x^2 - 6 = 0$

Giving $x^2 = 4$

$x = \pm 2$ (Note that you must include both solutions for $\sqrt{4}$)

Finding the second derivative:

> The second derivative is found by differentiating the first derivative.

$\dfrac{d^2y}{dx^2} = 3x$

When $x = 2$, $\dfrac{d^2y}{dx^2} = 3 \times 2 = 6$. This is a positive value indicating that the stationary point at

$x = 2$ is a minimum.

When $x = -2$, $\dfrac{d^2y}{dx^2} = 3 \times (-2) = -6$. This is a negative value indicating that the stationary point at $x = -2$ is a maximum.

To find the y-coordinate for each x-coordinate of the stationary points involves substituting the x-coordinate into the equation of the curve.

When $x = 2$, $y = \dfrac{1}{2} \times 8 - 6 \times 2 + 3 = -5$

When $x = -2$, $y = \dfrac{1}{2} \times (-8) - 6 \times (-2) + 3 = 11$

Hence there is a maximum point at $(-2, 11)$ and a minimum point at $(2, -5)$.

Example

③ The curve C has equation $y = x^2 - 8x + 10$.

(a) The point P has coordinates $(3, -5)$ and lies on C. Find the equation of the normal to C at P. [5]

(b) The point Q lies on C and is such that the tangent to C at Q has equation $y = 4x + c$, where c is a constant. Find the coordinates of Q and the value of c. [4]

(WJEC C1 May 2010 Q3)

Answer

③ (a) Differentiating the equation of the curve to find the gradient gives

$\dfrac{dy}{dx} = 2x - 8$

At P $(3, -5)$ the gradient is found by substituting $x = 3$ into the expression for $\dfrac{dy}{dx}$.

Hence, $\dfrac{dy}{dx} = 2(3) - 8 = -2$

The tangent and normal are perpendicular to each other, so

$(-2)\, m = -1$ giving $m = \dfrac{1}{2}$.

> The product of the gradients of perpendicular lines is -1. (i.e. $m_1 m_2 = -1$)

Equation of the normal having gradient $\frac{1}{2}$ and passing through $P(3, -5)$ is

$$y - (-5) = \frac{1}{2}(x - 3)$$

$$y + 5 = \frac{1}{2}(x - 3)$$

$$2y + 10 = x - 3$$

$$x - 2y - 13 = 0$$

(b) Tangent has equation $y = 4x + c$

Gradient of the tangent $= 4$

> This equation is of the form $y = mx + c$ where m is the gradient of the straight line.

Gradient of curve $= \dfrac{dy}{dx} = 2x - 8$

Hence $2x - 8 = 4$, giving $x = 6$.

To find the y-coordinate of Q, substitute $x = 6$ into the equation of the curve.

$$y = x^2 - 8x + 10$$

$$y = 6^2 - 8(6) + 10 = 36 - 48 + 10 = -2$$

So the coordinates of Q are $(6, -2)$

As the point Q lies on the tangent, the coordinates of Q must satisfy the equation of the tangent. Substituting $x = 6$ and $y = -2$ into the equation of the tangent gives:

$$y = 4x + c$$

$$-2 = 4(6) + c$$

Giving $c = -26$

Simple curve sketching

To sketch a curve when you are given its equation, you need to determine the following:

- The coordinates of the stationary points on the curve and their nature (i.e. whether they are maxima, minima or points of inflection).
- The coordinates of the points where the curve intersects (i.e. crosses) the x-axis.
- The coordinates of the point(s) where the curve intersects the y-axis.

You have already come across how to find the coordinates and the nature of the stationary points.

To find where a curve cuts the x-axis you substitute $y = 0$ and then solve the resulting equation in x.

To find where a curve cuts the y-axis you substitute $x = 0$ into the equation of the curve.

Once you have all these coordinates you can plot them on a suitable set of axes. The graph does not have to be drawn accurately as it is only a sketch but you have to include the coordinates of the points and make sure that the curve is drawn smoothly.

The following example shows all these techniques.

Example

① A curve has the equation $y = x^2 - 2x - 3$

    (a) Find the coordinates and nature of the stationary point of C.   [4]

    (b) Sketch the curve $y = x^2 - 2x - 3$.   [4]

Answer

① (a) $y = x^2 - 2x - 3$

$$\frac{dy}{dx} = 2x - 2$$

At the stationary point $\frac{dy}{dx} = 0$

Hence, $2x - 2 = 0$

Solving for x gives $x = 1$

Substituting $x = 1$ into the equation of the curve to find the corresponding y-coordinate gives

$$y = 1^2 - 2(1) - 3 = -4$$

Hence the coordinates of the stationary point are $(1, -4)$

> You could be asked to answer a question such as this using a different method, e.g. completing the square. For information on completing the square, look back at Topic 2.

Differentiating again to find the nature of the stationary point:

$$\frac{d^2y}{dx^2} = 2$$

The second derivative is positive showing that $(1, -4)$ is a minimum point.

(b) To determine where the curve cuts the x-axis, substitute $y = 0$ into the equation of the curve.

$$0 = x^2 - 2x - 3$$

Factorising gives $(x - 3)(x + 1) = 0$

Solving gives $x = 3$ or -1

To determine where the curve cuts the y-axis, substitute $x = 0$ into the equation of the curve.

$$y = (0)^2 - 2(0) - 3 = -3$$

You now need to draw a set of axes making sure that all the important points you have just found can be shown.

Substitute numbers on each axis where the curve cuts and make sure you mark on the curve the coordinates of the stationary point.

Remember to mark both axes and to write the equation next to the curve.

The curve can now be sketched like this:

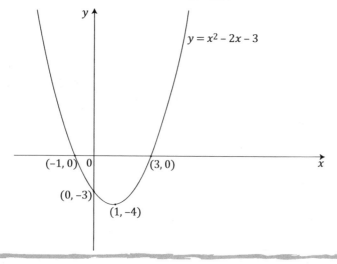

Examination style questions

① Given that $y = 2x^2 - 7x + 5$, show from first principles that

$$\frac{dy}{dx} = 4x - 7$$

[5]

Answer

Increasing x by a small amount δx will result in y increasing by a small amount δy.

Substituting $x + \delta x$ and $y + \delta y$ into the equation we have:

$y + \delta y = 2(x + \delta x)^2 - 7(x + \delta x) + 5$

$y + \delta y = 2(x^2 + 2x\delta x + (\delta x)^2) - 7x - 7\delta x + 5$

$y + \delta y = 2x^2 + 4x\delta x + 2(\delta x)^2 - 7x - 7\delta x + 5$

But $y = 2x^2 - 7x + 5$

Subtracting these equations gives

$\delta y = 4x\delta x + 2(\delta x)^2 - 7\delta x$

Dividing both sides by δx

$\frac{\delta y}{\delta x} = 4x + 2\delta x - 7$

Letting $\delta x \to 0$

$\frac{dy}{dx} = \underset{\delta x \to 0}{\text{limit}} \frac{\delta y}{\delta x} = 4x - 7$

② Differentiate $6x^{\frac{2}{3}} - \frac{3}{x^3}$ with respect to x. [2]

Answer

Writing in index form gives

$6x^{\frac{2}{3}} - 3x^{-3}$

Differentiating gives

$\left(\frac{2}{3}\right)6x^{-\frac{1}{3}} - (-3)3x^{-4} = 4x^{-\frac{1}{3}} + 9x^{-4}$

③ The curve C has equation $y = x^2 - 8x + 6$. The point A has coordinates $(1, 2)$.

(a) Find the equation of the tangent to C at A. [4]

(b) Find the equation of the normal to C at the point A. [2]

Answer

(a) $y = x^2 - 8x + 6$

Differentiating the equation of the curve to find the gradient gives

$\frac{dy}{dx} = 2x - 8$

At A (1, 2) the gradient is found by substituting $x = 1$ into the expression for $\dfrac{dy}{dx}$.

Hence, the gradient of the tangent at A, $\dfrac{dy}{dx} = 2(1) - 8 = -6$

The equation of the tangent is found using the formula:

$y - y_1 = m(x - x_1)$ where $m = -6$ and $(x_1, y_1) = (1, 2)$, so

$y - 2 = -6\,(x - 1)$

$y - 2 = -6x + 6$

$6x + y - 8 = 0$

(b) The tangent and normal are perpendicular to each other, so using $m_1 m_2 = -1$ we have

$(-6)\,m = -1$ giving $m = \dfrac{1}{6}$.

Equation of the normal having gradient $\dfrac{1}{6}$ and passing through $A(1, 2)$ is

$y - 2 = \dfrac{1}{6}(x - 1)$

$6y - 12 = x - 1$

$x - 6y + 11 = 0$

Test yourself

Answer the following questions and check your answers before moving onto the next topic.

① (a) Given that $y = 4x^2 + 2x - 1$ find $\dfrac{dy}{dx}$ from first principles.

(b) Given that $y = \dfrac{8}{x^2} + 5\sqrt{x} + 1$, find the gradient of the curve where $x = 1$.

② The curve C has the following equation

$y = 4\sqrt{x} + \dfrac{32}{x} - 3$

(a) Find the value of $\dfrac{dy}{dx}$ when $x = 4$.

(b) Find the equation of the normal to C at the point where $x = 4$.

③ The curve C has equation: $y = \dfrac{2}{3}x^3 + \dfrac{1}{2}x^2 - 6x$

Find the coordinates of the stationary points of C and determine the nature of these points.

④ A function is given by $f(x) = \sqrt{x^3} + 2x + 5$

Determine whether $f(x)$ is an increasing or decreasing function when $x = 4$.

⑤ A sheep pen is to be created using 100 m of fencing.

(a) Letting x be the length of this pen, find the length and width of the pen which would make the area of the pen a maximum.

(b) Find the area of the resulting pen.

(Note: answers to Test yourself are found at the back of the book.)

Q&A 1

1   The curve C has equation: $y = x^3 - 3x^2 + 3x + 5$

(a)   Show that C has only one stationary point. Find the coordinates of this point.   [4]

(b)   Verify that this stationary point is a point of inflection.   [2]

(WJEC C1 May 2009 Q10)

1

Answer

1 (a) $y = x^3 - 3x^2 + 3x + 5$

$$\frac{dy}{dx} = 3x^2 - 6x + 3 = 3(x^2 - 2x + 1) = 3(x - 1)(x - 1) = 3(x - 1)^2$$

At the stationary points $\frac{dy}{dx} = 0$

$3(x - 1)^2 = 0$

Solving gives $x = 1$ so there is only one stationary point.

To determine the y-coordinate of the stationary point, we substitute $x = 1$ into the equation of the curve.

$y = 1^3 - 3(1)^2 + 3(1) + 5 = 6$

So, the stationary point of curve C is at $(1, 6)$

(b)   Looking at the gradient either side of the stationary point at $x = 1$

When $x = 2$, $\frac{dy}{dx} = 3(2)^2 - 6(2) + 3 = 3$

When $x = 0$, $\frac{dy}{dx} = 3(0)^2 - 6(0) + 3 = 3$

The gradient does not change sign either side of the stationary point thus proving that the stationary point is a point of inflection.

Q&A 2

2   (a)   Given that $y = 3x^2 - 7x - 5$, find $\frac{dy}{dx}$ from first principles.   [5]

(b)   Given that $y = ax^{\frac{5}{2}}$, and $\frac{dy}{dx} = -2$ when $x = 4$, find the value of the constant a.   [3]

(WJEC C1 Jan 2010 Q6)

Answer

2 (a) $y = 3x^2 - 7x - 5$

Increasing x by a small amount δx will result in y increasing by a small amount δy.

Substituting $x + \delta x$ and $y + \delta y$ into the equation we have:

$y + \delta y = 3(x + \delta x)^2 - 7(x + \delta x) - 5$

$y + \delta x = 3(x^2 + 2x\delta x + (\delta x)^2) - 7x - 7\delta x - 5$

$y + \delta y = 3x^2 + 6x\delta x + 3(\delta x)^2 - 7x - 7\delta x - 5$

But $y = 3x^2 - 7x - 5$

Subtracting these equations gives

$\delta y = 6x\delta x + 3(\delta x)^2 - 7\delta x$

Dividing both sides by δx

$\dfrac{\delta y}{\delta x} = 6x + 3\delta x - 7$

Letting $\delta x \to 0$

$\dfrac{dy}{dx} = \underset{\delta x \to 0}{\text{limit}}\ \dfrac{\delta y}{\delta x} = 6x - 7$

(b) $y = ax^{\frac{5}{2}}$

> Remember: to differentiate you multiply by the index and then reduce the index by 1.

$\dfrac{dy}{dx} = \dfrac{5}{2}ax^{\frac{3}{2}}$

$\dfrac{dy}{dx} = \dfrac{5}{2}a\sqrt{x^3}$

> The denominator (i.e. bottom part) of a fractional power means a root (the 2 here means a square root). The numerator in the fractional power means the power to which the number is raised. If you are unsure about indices, then look back at Topic 1.

Now $\dfrac{dy}{dx} = -2$ when $x = 4$.

Hence $-2 = \dfrac{5}{2}a\sqrt{64}$

Solving for a gives $a = -\dfrac{1}{10}$

Summary C1 Pure Mathematics

Indices

Multiplying

$a^m \times a^n = a^{m+n}$

The indices are added together.

Dividing

$a^m \div a^n = a^{m-n}$

The indices are subtracted (i.e. top power minus bottom power).

Power raised to a power

$(a^m)^n = a^{m \times n}$

The indices are multiplied together.

Negative powers

$a^{-m} = \dfrac{1}{a^m}$

Zero power

If $a \neq 0$, $a^0 = 1$

Fractional powers

$a^{\frac{m}{n}} = \sqrt[n]{a^m} = \left(\sqrt[n]{a}\right)^m$

Negative fractional powers

$a^{-\frac{m}{n}} = \dfrac{1}{a^{\frac{m}{n}}} = \dfrac{1}{\sqrt[n]{a^m}} \text{ or } \dfrac{1}{\left(\sqrt[n]{a}\right)^m}$

Surds

Simple manipulation of surds

$\sqrt{a} \times \sqrt{a} = a$

$\sqrt{a} \times \sqrt{b} = \sqrt{ab}$

$(\sqrt{a} + \sqrt{b})(\sqrt{a} - \sqrt{b}) = a - b$

Rationalisation of surds

We avoid having surds in the denominator and removing them is called rationalising the denominator.

$$\frac{a}{b\sqrt{c}} = \frac{a}{b\sqrt{c}} \times \frac{\sqrt{c}}{\sqrt{c}} = \frac{a\sqrt{c}}{bc}$$

(Here the denominator is rationalised by multiplying the top and bottom by $\sqrt{c}$.)

$$\frac{a}{\sqrt{b} \pm \sqrt{c}} = \frac{a}{(\sqrt{b} \pm \sqrt{c})} \times \frac{(\sqrt{b} \mp \sqrt{c})}{(\sqrt{b} \mp \sqrt{c})} = \frac{a\sqrt{b} \mp a\sqrt{c}}{b - c}$$

(Here the denominator is rationalised by multiplying the top and bottom of the expression by the conjugate of the denominator.)

Quadratic functions and equations

Completing the square

The quadratic expression $ax^2 + bx + c$ can be written in the form $a(x + p)^2 + q$ and this is called completing the square. Completing the square can be used when a quadratic equation cannot be solved by factorisation or you want the find the maximum or minimum value of a quadratic function.

Solving/finding the roots of a quadratic equation when it cannot or cannot easily be factorised

$ax^2 + bx + c = 0$ has roots/solutions given by $x = \dfrac{-b \pm \sqrt{b^2 - 4ac}}{2a}$

Remember this formula as it will **not** be in the formula booklet.

Discriminants of quadratic functions

The discriminant of $ax^2 + bx + c$ is $b^2 - 4ac$.

For the equation $ax^2 + bx + c = 0$:

If $b^2 - 4ac > 0$, then there are two real and distinct (i.e. different) roots.

If $b^2 - 4ac = 0$, then there are two real and equal roots.

If $b^2 - 4ac < 0$, then there are no real roots.

Sketching a quadratic function

First write the equation $y = ax^2 + bx + c$ in the form $y = a(x + p)^2 + q$

From this equation:

If $a > 0$ the curve will be ∪-shaped.

If $a < 0$ the curve will be ∩-shaped.

The maximum or minimum point will be at $(-p, q)$.

The axis of symmetry will be $x = -p$

Solving linear inequalities

Solve them in the same way as you would solve ordinary equations but remember to reverse the inequality if you multiply or divide both sides by a negative quantity.

Solving quadratic inequalities

Consider the quadratic function equal to zero and solve to find the values of x where the curve cuts the x-axis.

Draw a sketch of the graph showing the intercepts on the x-axis.

If $ax^2 + bx + c < 0$ then the range of values of x covers the region below the x-axis.

If $ax^2 + bx + c > 0$ then the range of values of x covers the region above the x-axis.

If the inequality includes an equals sign then the range of values of x will include the values where it cuts the x-axis.

Transformations of the graph of $y = f(x)$

A graph of $y = f(x)$ can be transformed into a new function using the rules shown in this table.

Original function	New function	Transformation
$y = f(x)$	$y = f(x) + a$	Translation of a units parallel to the y-axis. (i.e. translation of $\binom{0}{a}$)
	$y = f(x + a)$	Translation of a units to the left parallel to the x-axis. (i.e. translation of $\binom{-a}{0}$)
	$y = f(x - a)$	Translation of a units to the right parallel to the x-axis. (i.e. translation of $\binom{a}{0}$)
	$y = -f(x)$	A reflection in the x-axis
	$y = af(x)$	One-way stretch with scale factor a parallel to the y-axis.
	$y = f(ax)$	One-way stretch with scale factor $\frac{1}{a}$ parallel to the x-axis.

Original function	New function	Transformation
$y = f(x)$	$y = f(x) + a$	

Original function	New function	Transformation
	$y = f(x + a)$	
	$y = f(x - a)$	
	$y = -f(x)$	
	$y = af(x)$ E.g. $y = 2\,f(x)$	
	$y = f(ax)$ E.g. $y = f(2x)$	

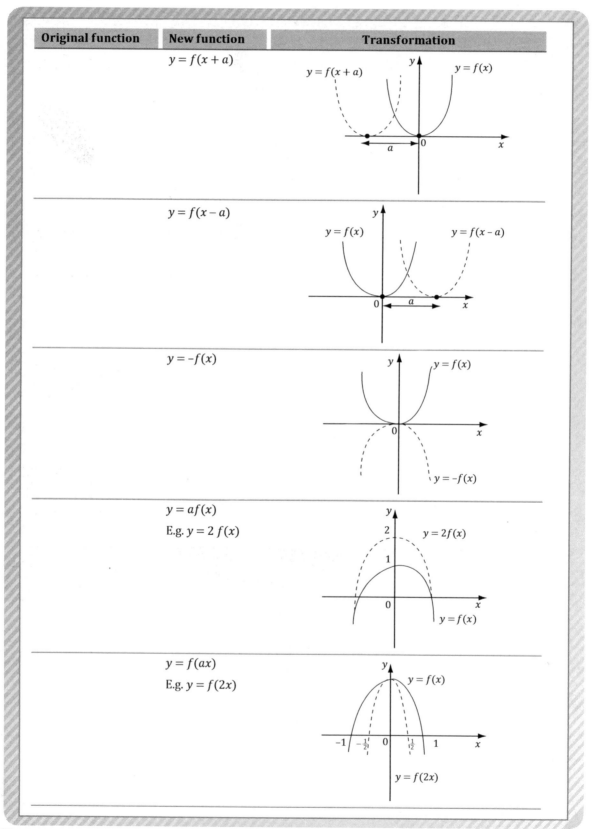

Co-ordinate geometry and straight lines

The gradient of the line joining two points

The gradient of the line joining points (x_1, y_1) and (x_2, y_2) is given by:

Gradient $= \dfrac{y_2 - y_1}{x_2 - x_1}$

The length of a line joining two points

The length of a straight line joining the two points (x_1, y_1) and (x_2, y_2) is given by:

$\sqrt{(x_2 - x_1)^2 + (y_2 - y_1)^2}$

The mid-point of the line joining two points

The mid-point of a line joining the points (x_1, y_1) and (x_2, y_2) is given by:

$\left(\dfrac{x_1 + x_2}{2}, \dfrac{y_1 + y_2}{2} \right)$

The equation of a straight line

The equation of a straight line with gradient m and which passes through a point (x_1, y_1) is given by:

$y - y_1 = m(x - x_1)$

Condition for two straight lines to be parallel to each other

The lines must both have the same gradient.

Condition for two straight lines to be perpendicular to each other

If one line has a gradient of m_1 and the other a gradient of m_2 then the lines are perpendicular to each other if $m_1 m_2 = -1$.

Polynomials and the binomial expansion

The remainder theorem

The remainder theorem states:

If a polynomial $f(x)$ is divided by $(x-a)$ the remainder is $f(a)$.

The factor theorem

For a polynomial $f(x)$, if $f(a) = 0$ then $(x-a)$ is a factor of $f(x)$.

The binomial expansion of $(a + b)^n$ for positive integer n

$$(a+b)^n = a^n + \binom{n}{1}a^{n-1}b + \binom{n}{2}a^{n-2}b^2 + \ldots + \binom{n}{r}a^{n-r}b^r + \ldots + b^n$$

$$\binom{n}{r} = {}^nC_r = \frac{n!}{r!(n-r)!}$$

The binomial expansion of $(1 + x)^n$ for positive integer n

$$(1+x)^n = 1 + nx + \frac{n(n-1)}{2!}x^2 + \frac{n(n-1)(n-2)}{3!}x^3 + \ldots + \frac{n(n-1)\ldots(n-r+1)}{r!}x^2 + \ldots$$

Differentiation

Differentiating

To differentiate terms of a polynomial expression: multiply by the index and then reduce the index by one.

If $y = kx^n$ then the derivative $\dfrac{dy}{dx} = nkx^{n-1}$

Increasing or decreasing functions

To find whether a curve or function is increasing or decreasing at a certain point:

Differentiate the equation of the curve or the function.

Substitute the value of x at the given point into the expression for the derivative to see whether the gradient is positive or negative. If the value is positive, the function is increasing at the given point, and if the value is negative, the function is decreasing.

Finding a stationary point

Substitute $\dfrac{dy}{dx} = 0$ and solve the resulting equation to find the value or values of x at stationary points.

Substitute the value or values of x into the equation of the curve to find the corresponding y-coordinate(s).

Finding whether a stationary point is a maximum or minimum

Differentiate the first derivative (i.e. $\dfrac{dy}{dx}$) to find the second derivative (i.e. $\dfrac{d^2y}{dx^2}$).

Substitute the x-coordinate of the stationary point into the expression for $\dfrac{d^2y}{dx^2}$.

If the resulting value is negative then the stationary point is a maximum point and if the resulting value is positive, then the stationary point is a minimum point. If $\dfrac{d^2y}{dx^2} = 0$, then the result is inconclusive and further investigation is required.

Determining whether a stationary point is a point of inflection

Substitute the x-coordinate of a point either side of the stationary point into the expression for $\dfrac{dy}{dx}$

and if the gradient has the same sign then the stationary point is a point of inflection.

Curve sketching

Find the points of intersection with the x and y axes by substituting $y = 0$ and $x = 0$ in turn and then solving the resulting equations.

Find the stationary points and their nature (i.e. maximum, minimum, point of inflection).

Plot the above on a set of axes.

It is important to note that maxima and minima are local maximum and minimum points only, and not necessarily the maximum or minimum values for a function. For example, the graph of a cubic equation shows this clearly.

Unit C2 Pure Mathematics 2

Unit C2 covers Pure Mathematics and seeks to build on your knowledge obtained from your C1 studies, so you may need to look back at this work. You must be proficient in the use of mathematical theories and techniques such as solving simple linear and quadratic equations, transposing formulae, algebraic manipulation, etc., and this may require you looking back over your GCSE work.

The knowledge, skills and understanding of the material in C2 will be built on in the other Pure Mathematics units as well as the units in mechanics or statistics which you will study. It will be assumed when you complete these other units that you have a thorough knowledge of the material covered in C1 and C2.

Revision checklist

Tick column 1 when you have completed all the notes.
Tick column 2 when you think you have a good grasp of the topic.
Tick column 3 during the final revision when you feel you have mastery of the topic.

		1	2	3	Notes
	1 Sequences, arithmetic series and geometric series				
p89	Arithmetic sequences and series				
p91	The summation sign and its use				
p93	Geometric sequences and series				
	2 Logarithms and their uses				
p101	$y = a^x$ and its graph				
p102	Logarithms and their proofs				
p104	The solutions of equations in the form $a^x = b$				
	3 Coordinate geometry of the circle				
p111	The equation of a circle				
p112	Circle properties				
p113	Finding the equation of a tangent to a circle				
p114	Finding where a circle and straight line intersect or meet				
p115	Using the discriminant to identify or show whether a line and circle intersect and, if so, how many times?				
	4 Trigonometry				
p124	Sine, cosine and tangent functions and their exact values				
p126	Obtaining angles given a trigonometric ratio				
p128	The sine and cosine rules				
p129	The area of a triangle				
p132	Radian measure, arc length, area of sector and area of segment				
p135	Sine, cosine and tangent graphs and their periodicity				

		1	2	3	Notes
p137	Knowledge and use of $\tan \theta = \dfrac{\sin\theta}{\cos\theta}$ and $\cos^2 \theta + \sin^2 \theta = 1$				
p137	Solution of simple trigonometric equations in a given interval				
	5 Integration				
p146	Indefinite integration as the reverse process of differentiation				
p148	Approximation of the area under a curve using the trapezium rule				
p149	Overestimating and underestimating areas using the trapezium rule				
p151	Interpretation of the definite integral as the area under a curve				

Topic 1 — Sequences, arithmetic series and geometric series

This topic covers the following:

- Sequences
- Arithmetic series
- Geometric series

Arithmetic sequences and series

In an arithmetic sequence, successive terms have a common difference d between them.

Take the following sequence, for example, 2, 5, 8, 11 ...

The above sequence has a first term of 2 and a common difference of 3. This common difference can be found by taking any term from the second term onwards and then subtracting the preceding term.

If the sequence starts with the first term a, then it carries on like this:

$$a, \qquad a + d, \qquad a + 2d, \qquad a + 3d \dots$$

Term: 1st   2nd   3rd   4th

The terms of an arithmetic sequence can be written in the following way:

$t_1 = a, t_2 = a + d, t_3 = a + 2d$, etc.

From the pattern in the terms you can see that the nth term, $t_n = a + (n - 1)d$.

Proof of the formula for the sum of an arithmetic series

An arithmetic series is formed when the terms of an arithmetic sequence are added together.
The sum of n terms of an arithmetic series can be written as:

$$S_n = a + (a + d) + (a + 2d) + \dots + (a + (n - 1)d) \tag{1}$$

The above sum starts from the first term and adds successive terms until the last term.
Now reversing the sum of the series starting from the last term, which we can call l, gives:

$$S_n = l + (l - d) + (l - 2d) + \dots l - (n - 1)d \tag{2}$$

Adding (1) and (2) together gives:

$$2S_n = (a + l) + (a + l) + (a + l) + \dots (a + l)$$

Notice that in the above, the $(a + l)$ appears n times.
Hence, we can write:

$$2S_n = n(a + l)$$

$$S_n = \frac{n}{2}(a + l)$$

The last term l can be written as $l = a + (n - 1)d$

> You must remember this proof as you may be asked to prove the formula $S_n = \frac{n}{2}[2a + (n - 1)d]$ in the examination.
> The formula for the sum of an arithmetic series, $S_n = \frac{n}{2}(a + l)$ can be a useful version of this formula at times.

So $S_n = \dfrac{n}{2}(a + a + (n-1)d)$

$$S_n = \dfrac{n}{2}\left[2a + (n-1)d\right]$$

The above formula appears in the formula booklet.

Example

① Find the sum of the first 20 terms of the arithmetic series which starts

4 + 11 + 18 + 25 + . . .

Answer

① First term $a = 4$ and common difference $d = 11 - 4 = 7$

$$S_n = \dfrac{n}{2}\left[2a + (n-1)d\right]$$

| This formula can be obtained from the formula booklet. |

$$S_{20} = \dfrac{20}{2}\left[2 \times 4 + (20-1)7\right]$$

$$S_{20} = 1410$$

Example

② (a) An arithmetic series has first term a and common difference d. Prove that the sum of the first n terms of the series is given by $S_n = \dfrac{n}{2}\left[2a + (n-1)d\right]$ [3]

(b) The eighth term of an arithmetic series is 28. The sum of the first 20 terms of the series is 710. Find the first term and the common difference of the arithmetic series.   [5]

(c) The first term of another arithmetic series is –3 and fifteenth term is 67. Find the sum of the first fifteen terms of this arithmetic series.   [2]

(WJEC C2 Jan 2011 Q4)

Answer

② (a) See Proof of the formula for the sum of an arithmetic series for this answer on page 89.

(b) $t_n = a + (n-1)d$

$t_8 = a + 7d$

$28 = a + 7d$ (1)

| Many of the exam questions in this topic require you to create equations using the information given in the question and then solve them simultaneously to find a and d. |

$$S_n = \dfrac{n}{2}\left[2a + (n-1)d\right]$$

$$S_{20} = \dfrac{20}{2}\left[2a + 19d\right]$$

$710 = 10(2a + 19d)$

$$71 = 2a + 19d \qquad\qquad (2)$$

Both sides are divided by 10.

Equations (1) and (2) are solved simultaneously.

$$71 = 2a + 19d$$
$$56 = 2a + 14d$$

Equation (1) is multiplied by 2 before subtracting.

Subtracting $\quad 15 = 5d$

Giving $d = 3$

Substituting $d = 3$ into equation (1) gives

$$28 = a + 21$$

Giving $a = 7$

Hence common difference $= 3$ and first term $= 7$

(c) $\quad t_n = a + (n - 1)d$

15th term $= -3 + 14d$

$$67 = -3 + 14d$$

Solving gives $d = 5$

By using the formula in the form $S_n = \dfrac{n}{2}(a+l)$, the answer can be found more readily i.e. $S_{15} = \dfrac{15}{2}(-3+67) = 480$.

$$S_n = \frac{n}{2}\left[2a + (n-1)d\right]$$

$$S_{15} = \frac{15}{2}\left[-6 + 14 \times 5\right]$$

$$= 480$$

The summation sign and its use

If you wanted to add the first four terms of an arithmetic sequence to form an arithmetic series, you can write it like this: $t_1 + t_2 + t_3 + t_4$

This can be written in the following way using a summation sign Σ.

$$\sum_{n=1}^{4} t_n$$

This means the sum of the terms from $n = 1$ to 4.

Take the following example. Here the terms are found by substituting $n = 1$, $n = 2$, $n = 3$, $n = 4$ and $n = 5$ into $(2n + 3)$. The terms are added together to form the series.

$$\sum_{n=1}^{5}(2n + 3) = 5 + 7 + 9 + 11 + 13 = 45$$

Example

① Evaluate

$$\sum_{n=1}^{3} n(n+1)$$

Answer

① $\displaystyle\sum_{n=1}^{3} n(n+1) = 1\times2 + 2\times3 + 3\times4 = 2+6+12 = 20$

Example

② Evaluate

$$\sum_{n=1}^{100}(2n-1)$$

Answer

② Series is $1 + 3 + 5 + 7 + 9\ldots$

> Start off by writing the first few terms so that a and d can be found.

First term $a = 1$ and common difference $d = 2$.

$$S_n = \frac{n}{2}\left[2a + (n-1)d\right]$$

> This formula is obtained from the formula booklet.

$$S_{100} = \frac{100}{2}\left[2\times1 + (100-1)2\right]$$

$$S_{100} = 50[2 + 99\times2]$$

> Remember to do the multiplication before the addition in the square bracket.

$$S_{100} = 10\,000$$

Example

③ The nth term of a number sequence is denoted by t_n.
The $(n + 1)^{\text{th}}$ term of the sequence satisfies $t_{n+1} = 2t_n - 3$
for all positive integers n and $t_4 = 33$.

(a) Evaluate t_1. [2]

(b) Explain why 40 098 cannot be one of the terms of this number sequence. [1]

Answer

③ (a) $t_{n+1} = 2t_n - 3$
   $t_4 = 2t_3 - 3$
   $33 = 2t_3 - 3$

> Work backwards by substituting t_4 into the equation to find t_3. Repeat by substituting t_3 in to find t_2. Finally substitute t_2 in to find the answer t_1.

   Solving gives $t_3 = 18$
   $t_3 = 2t_2 - 3$
   $18 = 2t_2 - 3$

Solving gives $t_2 = \dfrac{21}{2}$

$t_2 = 2t_1 - 3$

$\dfrac{21}{2} = 2t_1 - 3$

Solving gives $t_1 = \dfrac{27}{4}$

> Look carefully at the equation to see what would happen if t_n was odd or even.

(b) $t_{n+1} = 2t_n - 3$

Doubling t_n will always result in an even number when t_n is a whole number (i.e. from t_3 onwards).

Subtracting 3 from an even number always results in an odd number.

40 098 is even and therefore cannot be a term of the sequence.

Geometric sequences and series

Here is an example of a geometric sequence: $1, 5, 25, 125, \ldots$

From the second term onwards, if you divide one term by the term in front, you get the same number, which is called the common ratio.

In this series the common ratio is $\dfrac{25}{5} = 5$.

If the first term is a and the common ratio is r then a geometric sequence can be written as: $a, ar, ar^2, ar^3, \ldots$

Hence the first term $t_1 = a$, the second term $t_2 = ar$, the third term $t_3 = ar^2$, etc. Notice that the power of r is one less than the term number.

You can see that the nth term $t_n = ar^{n-1}$.

The common ratio is found by dividing the second term onwards by its preceding term.

Hence, $\dfrac{t_2}{t_1} = \dfrac{ar}{a} = r$, $\dfrac{t_3}{t_2} = \dfrac{ar^2}{ar} = r$, etc.

Proof of the formula for the sum of a geometric series

A geometric series is found by adding successive terms of a geometric sequence:

$a + ar + ar^2 + ar^3 + \ldots + ar^{n-1}$

The sum of n terms of a geometric series can be written as:

$S_n = a + ar + ar^2 + ar^3 + \ldots + ar^{n-1}$ \hfill (1)

Multiplying S_n by r gives

$rS_n = ar + ar^2 + ar^3 + \ldots + ar^n$ \hfill (2)

Subtracting equation (2) from equation (1) gives:

$S_n - rS_n = a - ar^n$

$S_n(1 - r) = a(1 - r^n)$

$S_n = \dfrac{a(1 - r^n)}{1 - r}$ provided $r \neq 1$

Example

① The fifth term of a geometric sequence is 96 and the eighth term is 768. Find the common ratio and the first term.

Answer

① $t_5 = ar^4 = 96$

$t_8 = ar^7 = 768$

Dividing these two terms $\dfrac{ar^7}{ar^4} = r^3 = \dfrac{768}{96} = 8$

Notice that dividing the terms, a cancels leaving an expression just in r.

Hence $r^3 = 8$ so common ratio $r = 2$

$ar^4 = 96$

So $a(2)^4 = 96$

Giving first term $a = 6$

The sum to infinity of a convergent geometric series

The following series is convergent: $1 + \dfrac{1}{2} + \dfrac{1}{4} + \dfrac{1}{8} + \ldots$

This means that successive terms are getting smaller in magnitude and that S_n approaches a certain limiting value. As $n \to \infty$, the sum of all the terms is called S_∞, the sum to infinity. The sum to infinity of a geometric series is given by:

$$S_\infty = \frac{a}{1-r} \text{ provided that } |r| < 1 \quad \left(S_n = \frac{a(1-r^n)}{1-r} \text{ provided } r \neq 1\right).$$

For a geometric series, $S_n = \dfrac{a(1-r^n)}{1-r} = \dfrac{a}{1-r} - \dfrac{ar^n}{1-r}$

If $|r| < 1$, r^n becomes very small as $n \to \infty$.

This means $\dfrac{ar^n}{1-r} \to 0$ as $n \to \infty$.

Hence $S_n \to \dfrac{a}{1-r}$ as $n \to \infty$.

If r takes a value not in the range $|r| < 1$ successive terms in the series become larger, so the series is divergent and a final limiting value is not reached. In this case S_∞ would not exist.

Example

① (a) Find the sum to infinity of the geometric series

$40 - 24 + 14 \cdot 4 - \ldots$ [3]

(b) Another geometric series has first term a and common ratio r. The fourth term of this geometric series is 8. The sum of the third, fourth and fifth terms of the series is 28.

(i) Show that r satisfies the equation: $2r^2 - 5r + 2 = 0$

(ii) Given that $|r| < 1$, find the value of r and the corresponding value of a. [6]

(WJEC C2 May 2010 Q6)

Answer

① (a) The terms of a geometric series are as follows:

$$a + ar + ar^2 + ar^3 + \ldots + ar^{n-1}$$

Common ratio $r = \dfrac{\text{2nd term}}{\text{1st term}}$

$$r = \frac{ar}{a} = -\frac{24}{40} = -\frac{3}{5}$$

$$S_\infty = \frac{a}{1-r}$$

$$= \frac{40}{1 - \left(-\dfrac{3}{5}\right)}$$

$$= 25$$

(b) (i) 4th term $= ar^3$

Hence $8 = ar^3$ so $a = \dfrac{8}{r^3}$

3rd term $= ar^2$

5th term $= ar^4$

Now $ar^2 + ar^3 + ar^4 = 28$

Substituting $a = \dfrac{8}{r^3}$ into this equation gives the following:

$$\frac{8}{r^3}r^2 + \frac{8}{r^3}r^3 + \frac{8}{r^3}r^4 = 28$$

$$\frac{8}{r} + 8 + 8r = 28$$

Multiplying both sides by r gives:

$$8 + 8r + 8r^2 = 28r$$
$$8r^2 - 20r + 8 = 0$$

Dividing through by 4 gives:

$$2r^2 - 5r + 2 = 0$$

(ii) Factorising $2r^2 - 5r + 2 = 0$

$$(2r - 1)(r - 2) = 0$$

Solving gives $r = \dfrac{1}{2}$ or $r = 2$

It is given that $|r| < 1$

So $r = \dfrac{1}{2}$

$$a = \frac{8}{r^3} = \frac{8}{\left(\dfrac{1}{2}\right)^3} = \frac{8}{\dfrac{1}{8}} = 64$$

> The nth term of a geometric series $= ar^{n-1}$.

Example

② A geometric series has first term a and common ratio r. The sum of the first two terms of the geometric series is 7.2. The sum to infinity of the series is 20. Given that r is positive, find the values of r and a. [6]

(WJEC C2 May 2008 Q5)

Answer

② 1st term $= a$, 2nd term $= ar$,

So $a + ar = 7.2$

$a(1 + r) = 7.2$

$S_\infty = \dfrac{a}{1-r}$

> This formula for the sum to infinity can be obtained from the formula booklet.

$20 = \dfrac{a}{1-r}$

$a = 20(1 - r)$

Substituting $a = 20(1 - r)$ into $a(1 + r) = 7.2$

This gives $20(1 - r)(1 + r) = 7.2$

$20(1 - r^2) = 7.2$
$r^2 = 0.64$
$r = \pm 0.8$ but r is positive, so $r = 0.8$
Now $a(1 + r) = 7.2$
$a(1 + 0.8) = 7.2$
$a = 4$

Example

③ (a) The second term of a geometric series is 6 and the fifth term is 384.
   (i) Find the common ratio of the series.
   (ii) Find the sum of the first eight terms of the geometric series. [6]

(b) The first term of another geometric series is 5 and the common ratio is 1.1.
   (i) The nth term of this series is 170, correct to the nearest integer. Find the value of n.
   (ii) Dafydd, who has been using his calculator to investigate various properties of this geometric series, claims that the sum to infinity of the series is 940. Explain why this result cannot possibly be correct. [5]

(WJEC C2 Jan 2011 Q5)

Answer

③ (a)(i) $t_2 = ar$
         So, $6 = ar$ (1)
         $t_5 = ar^4$
         So, $384 = ar^4$ (2)

Dividing equation (2) by equation (1) gives

$$\frac{384}{6} = \frac{ar^4}{ar}$$

$$64 = r^3$$

Giving $r = 4$

(ii) Substituting $r = 4$ into equation (1)

$$6 = 4a$$

So $a = \frac{3}{2}$

> It is often easier to write the formula in the form $S_n = \frac{a(r^n - 1)}{r - 1}$ when $r > 1$, thus avoiding a negative numerator and denominator.

$$S_n = \frac{a(1 - r^n)}{1 - r}$$

> This formula can be obtained from the formula booklet.

$$S_8 = \frac{\frac{3}{2}(1 - 4^8)}{1 - 4}$$

> Remember that you are allowed to use a calculator in the examination for C2.

$$= 32\,767.5$$

(b)(i) $t_n = ar^{n-1}$

Hence $170 = 5\,(1.1)^{n-1}$

> Equations involving powers like this are solved by taking logs of both sides.

$$34 = (1.1)^{n-1}$$

Taking $\log_{10}$ of both sides: $\log_{10}34 = \log_{10}(1.1)^{n-1}$

$$\log_{10}34 = (n - 1)\log_{10}1.1$$

$$\frac{\log_{10}34}{\log_{10}1.1} = n - 1$$

> See the next topic if you are unsure about taking logs of both sides.

$$36.9988 = n - 1$$

$$n = 37.9988$$

n has to be an integer, so $n = 38$

(ii) The common ratio is 1.1

For a sum to infinity to exist $|r| < 1$ so in this case the sum to infinity does not exist.

Grade boost

Always look back at the question to see if there are any conditions placed on the value you have found. Here it has to be an integer.

Examination style questions

① In an arithmetic series the ninth term is double the fourth term. If the sixteenth term is 68, find the first term and the common difference of this arithmetic series. [5]

Answer

① $t_n = a + (n - 1)d$

$t_{16} = a + 15d = 68$ (1)

$t_9 = a + 8d$

$t_4 = a + 3d$

> It is best to number simultaneous equations so they can be referred to by their number.

Now ninth term is double the fourth term, so

$a + 8d = 2(a + 3d)$

$a = 2d$

Substituting $a = 2d$ into equation (1) gives

$2d + 15d = 68$

$17d = 68$

$d = 4$

$a = 2d = 8$

Hence first term $a = 8$ and common difference $d = 4$.

> You can write this as follows: $t_9 = 2t_4$

> It is easy to make a mistake when solving simultaneous equations, so remember to check them by substituting both values into the equation which has not been used for the substitution. If the right-hand side equals the left-hand side, your values are likely to be correct.

② The sixth term of a geometric sequence is 2187 and the fourth term is 243.
If the common ratio is positive, find the common ratio and the first term.

Answer

② $t_6 = ar^5 = 2187$

$t_4 = ar^3 = 243$

$\dfrac{t_6}{t_4} = \dfrac{ar^5}{ar^3} = r^2 = \dfrac{2187}{243} = 9$

$r = \pm 3$ but r is positive

Hence $r = 3$

$ar^3 = 243$

$a \times 27 = 243$

$a = 9$

Hence first term $a = 9$ and common ratio $r = 3$

> Look at the question carefully to see if it is possible to have both values.

Test yourself

Answer the following questions and check your answers before moving on to the next topic.

① Find an expression, in terms of n, for the sum of the first n terms of the arithmetic series:
$4 + 10 + 16 + 22 + \ldots$

Simplify your answer.

② The sum of the first seven terms of an arithmetic series is 182. The sum of the fifth and seventh terms of the series is 80. Find the first term and the common difference of the series.

③ A geometric series has first term a and common ratio r. The sum of the first two terms of the geometric series is 2.7. The sum to infinity of the series is 3.6. Given that r is positive, find the values of r and a.

(Note: answers to Test yourself are found at the back of the book.)

1 (a) An arithmetic series has first term a and common difference d.

Prove that the sum of the first n terms of the series is given by

$$S_n = \frac{n}{2}\left[2a+(n-1)d\right]$$ [3]

(b) The first term of an arithmetic series is 4 and the common difference is 2.

The sum of the first n terms of the arithmetic series is 460.

Write down an equation satisfied by n. Hence find the value of n. [3]

(c) The fifth term of another arithmetic series is 9. The sum of the sixth term and the tenth term of this series is 42. Find the first term and the common difference of the arithmetic series. [5]

(WJEC C2 May 2010 Q5)

Answer

1 (a) See section on 'Proof of the formula for the sum of an arithmetic series' on page 89.

(b) $a = 4$ and $d = 2$.

$S_n = 460$

$$S_n = \frac{n}{2}\left[2a+(n-1)d\right]$$

$$460 = \frac{n}{2}\left[8+(n-1)2\right]$$

> Substituting $a = 4$, $d = 2$ and $S_n = 460$ into the formula for S_n.

$920 = n(2n + 6)$

$920 = 2n^2 + 6n$

$460 = n^2 + 3n$

> Divide both sides by 2.

$n^2 + 3n - 460 = 0$

$(n + 23)(n - 20) = 0$

> This is quite a hard quadratic to factorise. The two factors need to be close together to give $+3n$ in the middle.

$n = 20$ as the other value would mean a negative number of terms.

(c) $t_5 = a + 4d$

$9 = a + 4d$ ⠀⠀⠀⠀⠀(1)

$t_6 = a + 5d$

$t_{10} = a + 9d$

$t_6 + t_{10} = a + 5d + a + 9d = 2a + 14d$

$2a + 14d = 42$

$21 = a + 7d$ ⠀⠀⠀⠀⠀(2)

Solving equations (1) and (2) simultaneously

Equation (2) − equation (1) gives

$3d = 12$

$d = 4$

Substituting this value of d into equation (1)

$9 = a + 4 \times 4$

$9 = a + 16$

$a = -7$

Hence first term $a = -7$ and common difference $d = 4$

2   The nth term of a number sequence is denoted by t_n. The $(n + 1)$th term of the sequence satisfies: $t_{n+1} = 2t_n + 1$ for all positive integers n.

Given that $t_4 = 63$,

(a)   evaluate t_1,   [2]

(b)   without carrying out any further calculations, explain why 6 043 582 cannot be one of the terms of this number sequence.   [1]

(WJEC C2 Jan 2010 Q10)

Answer

2 (a) $t_{n+1} = 2t_n + 1$

$t_4 = 2t_3 + 1$

$63 = 2t_3 + 1$

$t_3 = 31$

$t_3 = 2t_2 + 1$

$31 = 2t_2 + 1$

$t_2 = 15$

$t_2 = 2t_1 + 1$

$15 = 2t_1 + 1$

$t_1 = 7$

(b) 6 043 582 is even but all the terms of the sequence are odd.

2 × (an even or odd number) always results in an even number and adding a one to an even number will make an odd number.

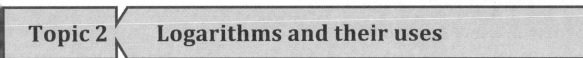

Topic 2 — Logarithms and their uses

This topic covers the following:

- $y = a^x$ and its graph
- Laws of logarithms
- The solutions of equations in the form $a^x = b$

$y = a^x$ and its graph

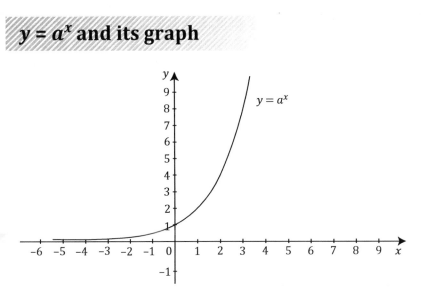

The graph of $y = a^x$ where a is a positive constant greater than 1, is shown above. Notice that the graph intersects the y-axis at 1. No matter what the positive value of a is, the intersect on the y-axis is always 1. The reason for this is that on the y-axis, $x = 0$ so $y = a^0 = 1$.

If the value of the positive constant a is less than 1, then the graph of $y = a^x$ looks like this.

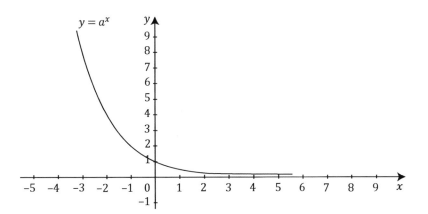

Logarithms and their proofs

What is a logarithm?

A logarithm of a positive number to a base a is the power to which the base must be raised in order to give the positive number. We can write two important equations using this definition where the positive number is y, the base is a and the log of the number is x:

$y = a^x$
$\log_a y = x$

> Both of these equations have the same meaning and you must be able to convert between them.

The following example will help explain this:

If $y = 10^3$ then from the second equation we have $\log_{10} y = 3$

> You must remember both of these equations and be able to use them.

For a positive base a, the following are true:

$\log_a a = 1$, as $a^1 = a$
$\log_a 1 = 0$, as $a^0 = 1$

Proof of the laws of logarithms

The following proofs of the laws of logarithms need to be remembered for the examination. You will also be required to use these laws.

Proving $\log_a x + \log_a y = \log_a (xy)$

Suppose $x = a^n$ and $y = a^m$

These can be rewritten as:

$\log_a x = n$ and $\log_a y = m$

$xy = a^n \times a^m$

$xy = a^{n+m}$

> Remember that according to the laws of indices the powers here are added.

This can be rewritten as:

$\log_a (xy) = n + m$

But $n = \log_a x$ and $m = \log_a y$

$\log_a x + \log_a y = \log_a (xy)$

Proving $\log_a x - \log_a y = \log_a (x/y)$

Suppose $x = a^n$ and $y = a^m$

These can be rewritten as:

$\log_a x = n$ and $\log_a y = m$

Now $\dfrac{x}{y} = \dfrac{a^n}{a^m}$

$$\frac{x}{y} = a^{n-m}$$

> Remember that according to the laws of indices the powers here are subtracted.

This can be rewritten as:

$$\log_a \frac{x}{y} = n - m$$

But $n = \log_a x$ and $m = \log_a y$

$$\log_a \frac{x}{y} = \log_a x - \log_a y$$

Hence

$$\log_a x - \log_a y = \log_a \frac{x}{y}$$

Proving $k \log_a x = \log_a (x^k)$

Suppose $x = a^n$, then $\log_a x = n$

Raising both sides of the equation to the power k gives

$$x^k = (a^n)^k$$

$$x^k = a^{nk}$$

> According to the laws of indices, you multiply the powers inside and outside the brackets.

This can be rewritten as:

$$\log_a x^k = nk$$

As $n = \log_a x$ this can be substituted in for n.

Hence

$$\log_a x^k = k \log_a x$$

All three laws can be used when dealing with expressions or equations involving logs. Examples involving the simplification of expressions containing logs are shown here.

Example

① Express $\log_a 64 - 2 \log_a 4$ as a single logarithm in the form $\log_a b$ where b is an integer.

Answer

① $\log_a 64 - 2 \log_a 4$

$= \log_a 64 - \log_a 4^2$

> Here we are using this law of logarithms $\log_a x^k = k \log_a x$

$= \log_a 64 - \log_a 16$

$= \log_a \frac{64}{16}$

> Here we are using this law of logarithms $\log_a x - \log_a y = \log_a \frac{x}{y}$

$= \log_a 4$

Example

② Express $\frac{1}{2}\log_a 9 + \log_a 3 - 3\log_a 3$ as a single term.

Answer

② $\frac{1}{2}\log_a 9 + \log_a 3 - 3\log_a 3$

$= \log_a 9^{\frac{1}{2}} + \log_a 3 - \log_a 3^3$

$= \log_a 3 + \log_a 3 - \log_a 27$

$= \log_a\left(\frac{3 \times 3}{27}\right)$

$= \log_a \frac{1}{3}$

$= \log_a 1 - \log_a 3$

$= -\log_a 3$

> Note that $9^{\frac{1}{2}} = \sqrt{9} = \pm 3$. As you cannot have the logarithim of a negative number, only the positive value is used.

> Use $\log_a \frac{1}{x} = -\log_a x$

The solution of equations in the form $a^x = b$

Equations in the form $a^x = b$ can be solved by first taking logs to base a of both sides like this:

$a^x = b$

$x = \log_a b$

> $\log_a a^x = x \log_a a = x$
> as $\log_a a = 1$

Alternatively, you should recognise these two equations as having the same meaning.

Example

① Solve the equation $\log_4 x = -\frac{1}{2}$

Answer

① $\log_4 x = -\frac{1}{2}$

$x = 4^{-\frac{1}{2}}$

$x = \frac{1}{4^{\frac{1}{2}}}$

$x = \frac{1}{\sqrt{4}}$

$x = \pm\frac{1}{2}$

x cannot be negative so $x = \frac{1}{2}$

> **Grade boost**
>
> Ensure you are fully confident in converting from log to exponential form and vice versa.

> When taking a square root, you must remember to include the $\pm$; however, you cannot find the logarithm of a negative number so the negative solution is ignored here.

Example

② Solve the equation $2^{2x-1} = 9$
giving your answer correct to three decimal places.

Answer

② $2^{2x-1} = 9$

$\log 2^{2x-1} = \log 9$

$(2x - 1) \log 2 = \log 9$

$2x - 1 = \dfrac{\log 9}{\log 2}$

$2x - 1 = 3.1699$

$2x = 4.1699$

$x = 2.085$ (3 d.p.)

> Questions like this are solved by taking logarithms of both sides. The base used here is base 10. If the base is not shown, it is assumed that it is base 10.

> Use the Log button on your calculator to work out the logs of numbers. Note that here $\dfrac{\log 9}{\log 2}$ does **not** equal $\log \dfrac{9}{2}$.

> When giving a numerical answer always check whether the answer needs to be given to a certain number of significant figures or decimal places.

Example

③ Solve the equation $\log_a (3x + 4) = \log_a 5 + \log_a x$

Answer

③ $\log_a (3x + 4) = \log_a 5 + \log_a x$
$\log_a (3x + 4) - \log_a x = \log_a 5$

$\log_a \left(\dfrac{3x + 4}{x} \right) = \log_a 5$

$\dfrac{3x + 4}{x} = 5$

$3x + 4 = 5x$

$4 = 2x$

$x = 2$

Example

④ Solve the equation $25^x - 4 \times 5^x + 3 = 0$ where $x > 0$

Answer

④ $25^x - 4 \times 5^x + 3 = 0$
Now $25^x = (5^2)^x = 5^{2x} = (5^x)^2$
Hence $5^{2x} - 4 \times 5^x + 3 = 0$
So $(5^x)^2 - 4 \times 5^x + 3 = 0$

Let $y = 5^x$
$y^2 - 4y + 3 = 0$
$(y - 1)(y - 3) = 0$

> You need to recognise that this equation is similar in format to a quadratic equation.

> Notice that $25^x = (5^2)^x = 5^{2x} = (5^x)^2$

> Notice that $25^x = 5^{2x}$. The substitution $y = 5^x$ is used to obtain a quadratic equation in y which can then be factorised and solved.

105

$y = 1$ or $y = 3$

When $y = 1$, $1 = 5^x$

$5^0 = 1$ so $x = 0$

We cannot have this value as x must be greater than 0.

When $y = 3$

$5^x = 3$

Taking logs to base 10 of both sides

$\log 5^x = \log 3$

$x \log 5 = \log 3$

$x = \dfrac{\log 3}{\log 5}$

$x = 0.68$ (2 d.p.)

> When there is more than one value for x always check whether each value is possible. Look back at the question to see if there is a restriction on x. Here the restriction is $x > 0$

Example

⑤ (a) Given that $x > 0$ show that $\log_a x^n = n \log_a x$

(b) Express $\dfrac{1}{2}\log_a 324 + \log_a 56 - 2\log_a 12$ in the form $\log_a b$, where b is a constant whose value is to be found. [4]

(c) (i) Rewrite the equation $3^x = 2^{x+1}$

in the form $\qquad c^x = d$

where the values of the constants c and d are to be found.

(ii) Hence or otherwise, solve the equation $3^x = 2^{x+1}$ giving your answer correct to two decimal places. [4]

(WJEC C2 Jan 2010 Q7)

Answer

⑤ (a) See proof on page 103.

(b) $\dfrac{1}{2}\log_a 324 + \log_a 56 - 2\log_a 12$

> $324^{\frac{1}{2}} = \sqrt{324} = 18$

$= \log_a 324^{\frac{1}{2}} + \log_a 56 - \log_a 12^2$

$= \log_a 18 + \log_a 56 - \log_a 144$

$= \log_a\left(\dfrac{18 \times 56}{144}\right)$

$= \log_a 7$

(c) (i) $3^x = 2^{x+1}$

$3^x = 2^x \times 2^1$

> Notice the way the indices are separated here. The laws of indices are used here to separate 2^{x+1} into $2^x \times 2^1$.

$$\frac{3^x}{2^x} = 2$$

$$\left(\frac{3}{2}\right)^x = 2$$

Hence $c = \dfrac{3}{2}$ and $d = 2$

> Look back at the question to check if you have the answer in the correct format. Here we need an answer in the form $c^x = d$.
>
> The answer is requested in this format with $c = \dfrac{3}{2}$ and $d = 2$.

(ii) $\left(\dfrac{3}{2}\right)^x = 2$

Taking logs of both sides:

$$\log\left(\frac{3}{2}\right)^x = \log 2$$

> When you have an answer, always check whether it needs to be given to a certain number of decimal places or significant figures.

$$x = \frac{\log 2}{\log\left(\dfrac{3}{2}\right)} = \frac{0.3010}{0.1761} = 1.71 \ (2 \ \text{d.p.})$$

Examination style questions

① Solve $6^x = 12$, giving your answer correct to three decimal places. [3]

Answer

① $6^x = 12$

Taking logs of both sides

$\log 6^x = \log 12$

$x \log 6 = \log 12$

$x = \dfrac{\log 12}{\log 6}$

$x = 1.387$ (3 d.p.)

> The rule $\log_a x^k = k \log_a x$ is used here.

② Solve the equation $9^x - 6 \times 3^x + 8 = 0$ where $x > 0$
giving x correct to two decimal places. [5]

Answer

② $9^x - 6 \times 3^x + 8 = 0$

Now $9^x = (3^2)^x = 3^{2x} = (3^x)^2$

$(3^x)^2 - 6 \times 3^x + 8 = 0$

Let $y = 3^x$

$y^2 - 6y + 8 = 0$

$(y - 4)(y - 2) = 0$

Hence $y = 4$ or $y = 2$

When $y = 4$, $4 = 3^x$

Taking logs of both sides

$\log 4 = \log 3^x$

$\log 4 = x \log 3$

$x = \dfrac{\log 4}{\log 3}$

$x = 1.26$ (2 d.p.)

When $y = 2$, $2 = 3^x$

Taking logs of both sides

$\log 2 = \log 3^x$

$\log 2 = x \log 3$

$x = \dfrac{\log 2}{\log 3}$

$x = 0.63$ (2 d.p.)

$x = 1.26$ or 0.63 to 2 d.p.

> You need to recognise that this equation has the format of a quadratic equation. Notice that 9^x can be written as $(3^2)^x$ which can then be written as 3^{2x} and $(3^x)^2$.

> A substitution $y = 3^x$ is used here to obtain a quadratic equation in y. This will make the factorisation easier.

Test yourself

Answer the following questions and check your answers before moving on to the next topic.

① Simplify $\log_2 36 - 2\log_2 15 + \log_2 100$

expressing your answer in the form $\log_2 a$ where a is an integer.

② Solve the equation $\log_{27} x = \dfrac{2}{3}$

③ Solve the equation $3^x = 2$

giving your answer correct to two decimal places.

④ Express $\dfrac{1}{2}\log_a 36 - 2\log_a 6 + \log_a 4$ as a single logarithm.

⑤ Solve the equation $\log_a(6x^2 + 5) - \log_a x = \log_a 17$

(Note: answers to Test yourself are found at the back of the book.)

Q&A 1

1   Find all the values of x satisfying the equation $\log_a (6x^2 + 11) - \log_a x = 2 \log_a 5$ [5]

(WJEC C2 Jan 2011 Q7)

Answer

1    $\log_a (6x^2 + 11) - \log_a x = 2 \log_a 5$

$$\log_a\left(\frac{6x^2 + 11}{x}\right) = \log_a 5^2$$

$$\frac{6x^2 + 11}{x} = 25$$

$6x^2 + 11 = 25x$

$6x^2 - 25x + 11 = 0$

$(3x - 11)(2x - 1) = 0$

Hence $x = \dfrac{11}{3}$ or $x = \dfrac{1}{2}$

> Form a quadratic equation by getting all the terms onto the same side so they equal zero.

> ⤊ **Grade boost**
>
> Answers only with no working will earn 0 marks.

Q&A 2

2 (a)   Given that $x > 0$, show that $\log_a x^n = n \log_a x$ [3]

  (b)   Solve the equation $6^{2y - 1} = 4$

          Show your working and give your answer correct to three decimal places.   [3]

  (c)   Given that $\log_a 4 = \dfrac{1}{2}$, find the value of a.   [2]

(WJEC C2 May 2010 Q8)

Answer

2 (a)   See proof on page 103.

  (b) $6^{2y - 1} = 4$

          $\log 6^{2y - 1} = \log 4$

          $(2y - 1) \log 6 = \log 4$

          $2y - 1 = \dfrac{\log 4}{\log 6}$

          $2y - 1 = 0.7737$

          $2y = 1.7737$

          Solving gives $y = 0.887$ to 3 d.p.

  (c) $\log_a 4 = \dfrac{1}{2}$

          $a^{\frac{1}{2}} = 4$

          $a = 16$

> Working should always be carried out to at least one place more than the accuracy requested; e.g. if the question asks for the answer correct to 3 decimal places (as here), then working should be shown to at least 4 d.p. before rounding to 3 d.p. at the end.

> Here you are using the fact that $\log_a x^k = k \log_a x$

> The two equations $x = a^n$ and $\log_a x = n$ given in the formula booklet have the same meaning and you need to be able to convert between the logarithm and power representations readily.

> Square both sides to find a.

Topic 3	Coordinate geometry of the circle

This topic covers the following:

- The equation of the circle
- Circle properties
- Finding the equation of a tangent to a circle
- Finding where a circle and straight line intersect or meet
- The case for the circle and line not meeting or intersecting

The equation of a circle

The equation of a circle can be written in the form:

$$(x - a)^2 + (y - b)^2 = r^2$$

A circle having the above equation will have centre (a, b) and radius r.

There is the following alternative form for the equation of a circle:

$$x^2 + y^2 + 2gx + 2fy + c = 0$$

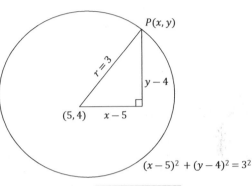

A circle having the above equation will have centre $(-g, -f)$ and radius given by $r = \sqrt{g^2 + f^2 - c}$

Note: You have to remember this alternative form for the equation as well as be able to work out the centre and radius. Remembering this is hard, so there is an alternative method which involves completing the square and this method is shown in Example 3.

Example

① Find the coordinates of the centre and the radius of the circle having the equation:

$$(x - 7)^2 + (x + 3)^2 = 36$$

Answer

① Comparing the equation $(x - 7)^2 + (x + 3)^2 = 36$ with the equation for the circle

$$(x - a)^2 + (y - b)^2 = r^2$$

This gives $a = 7$ and $b = -3$ so coordinates of the centre are $(7, -3)$.

$r^2 = 36$, giving radius $r = \sqrt{36} = 6$.

Example

② The circle C has centre A and equation $x^2 + y^2 - 2x + 6y - 6 = 0$.
Write down the coordinates of A and find the radius of C.

Answer

② Comparing the equation $x^2 + y^2 - 2x + 6y - 6 = 0$ with the equation

$x^2 + y^2 + 2gx + 2fy + c = 0$, we can see $g = -1$, $f = 3$ and $c = -6$.

Centre A has coordinates $(-g, -f) = (1, -3)$

Radius $= \sqrt{g^2 + f^2 - c} = \sqrt{(-1)^2 + (3)^2 + 6} = \sqrt{16} = 4$

Example

③ The circle C has centre A and equation $x^2 + y^2 - 4x + 2y - 11 = 0$

Find the coordinates of A and the radius of C.

Answer

③ The equation for C can be written as

$x^2 - 4x + y^2 + 2y - 11 = 0$

Completing the square means $x^2 - 4x = (x - 2)^2 - 4$

Similarly $y^2 + 2y = (y + 1)^2 - 1$

Hence, equation of C can be written as

> Completing the square is a good way of finding the centre and radius because you do not need to remember the formula involving f and g.

$(x - 2)^2 - 4 + (y + 1)^2 - 1 - 11 = 0$

$(x - 2)^2 + (y + 1)^2 - 4 - 1 - 11 = 0$

$(x - 2)^2 + (y + 1)^2 = 16$

Comparing this with the equation of the circle $(x - a)^2 + (y - b)^2 = r^2$

gives the coordinates of centre A as $(2, -1)$ and radius $= 4$.

Circle properties

There are a number of circle properties you need to know about.

① The angle in a semicircle is always a right angle.

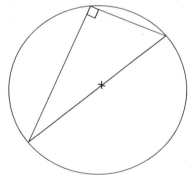

② The perpendicular from the centre of a circle to a chord bisects the chord.

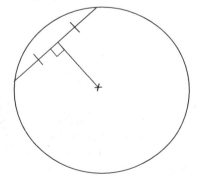

③ The tangent to a circle at a point makes a right angle with the radius of the circle at the same point.

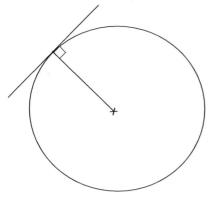

Finding the equation of a tangent to a circle

If you know the coordinates of the point where the tangent touches the circle and the coordinates of the centre of the circle, then you can find the gradient of the line joining these two points using the formula:

$$\text{Gradient} = \frac{y_2 - y_1}{x_2 - x_1}$$

≫ Grade boost

There are a number of formulae that come from Core 1 that you will need to remember and use. The formulae are listed in the summary.

You would then use this gradient to work out the gradient of the tangent as these two lines are perpendicular to each other. If one line has a gradient m_1 and the other a gradient of m_2 then as the lines are perpendicular $m_1 m_2 = -1$.

You would then use the coordinates of the point where the tangent touched the circle and the gradient of the tangent and substitute them into following formula to give the equation of the tangent.

$$y - y_1 = m(x - x_1)$$

The following example will help explain the method.

Example

① Circle C has centre A and equation $x^2 + y^2 - 4x + 2y - 20 = 0$.

    (a)   Find the coordinates of the centre A and the radius of C.   [3]

    (b)   The point P has coordinates $(5, 3)$ and lies on circle C. Find the equation of the tangent to C at P.   [4]

Answer

①   (a)   We will use the method of completing the square here to work out the coordinates of the centre A and the radius of the circle C.

        $x^2 + y^2 - 4x + 2y - 20 = 0$.

        $(x - 2)^2 + (y + 1)^2 - 4 - 1 - 20 = 0$

        $(x - 2)^2 + (y + 1)^2 = 25$

        $(x - 2)^2 + (y + 1)^2 = 5^2$

> Completing the square is used here but you could of course use the alternative method involving the formula. You would need to remember the formula and how to use it as it is not in the formula booklet. See example 2 on page 111.

        Hence coordinates of the centre A are $(2, -1)$ and radius is 5.

(b)   Gradient of the line joining the centre of the circle $A(2, -1)$ to point $P(5, 3)$ is given by:

$$\text{Gradient} = \frac{y_2 - y_1}{x_2 - x_1} = \frac{3 - (-1)}{5 - 2} = \frac{4}{3}$$

Line AP is a radius of the circle. The tangent at point P will be perpendicular to the radius AP.

For perpendicular lines, the product of the gradients $= -1$

Hence $m \times \left(\frac{4}{3}\right) = -1$

Gradient of tangent $m = -\frac{3}{4}$

Equation of the tangent having gradient $m = -\frac{3}{4}$ and passing through the point $P(5, 3)$ is

$$y - 3 = -\frac{3}{4}(x - 5)$$

$$4y - 12 = -3x + 15$$

$$3x + 4y - 27 = 0$$

> The formula for a straight line is used here. The formula for the equation of a straight line having gradient m and passing through the point (x_1, y_1) is $y - y_1 = m(x - x_1)$.

Finding where a circle and straight line intersect or meet

There are two ways in which a straight line can intersect or meet a circle:

① The line and circle can intersect in two places like this:

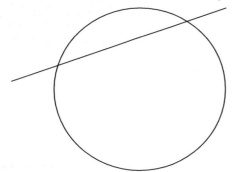

② The line and circle can meet in one place. This means the straight line becomes a tangent to the circle and also makes a right angle with the radius.

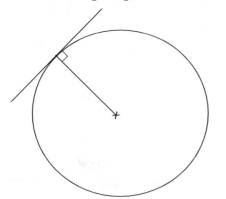

To find the coordinates of intersection or meeting

To find the coordinates you need to know the equation of the circle and the equation of the straight line. These are then solved simultaneously. You can use the straight line equation to find x in terms of y or y in terms of x. You then substitute this into the equation of the circle and then solve the resulting equation. Sometimes there will be two different roots (i.e. solutions), which means the circle and line cut in two places. Sometimes there will be two equal roots which means the circle and line meet in one place, i.e. the line is a tangent to the circle.

If there are no real roots to the equation, it means the line and circle do not intersect.

Example

① A circle C has equation $x^2 + y^2 + 2x - 12y + 12 = 0$.

The line with equation $x + y = 4$ intersects the circle at two points P and Q. Find the coordinates of P and Q.

Answer

① To find the points of intersection we solve the two equations simultaneously.

$x + y = 4$

So $y = 4 - x$

Substituting $y = 4 - x$ into the equation of the circle, gives

$x^2 + (4 - x)^2 + 2x - 12(4 - x) + 12 = 0$

$x^2 + 16 - 8x + x^2 + 2x - 48 + 12x + 12 = 0$

$2x^2 + 6x - 20 = 0$

Dividing by two gives $x^2 + 3x - 10 = 0$

> Always look at a quadratic to see if all the terms can be divided by the same number. This will make the factorisation easier.

Factorising gives $(x + 5)(x - 2) = 0$

Solving gives $x = -5$ or 2

These two x-coordinates are substituted into the equation of the line to find the corresponding y-coordinates.

When $x = -5$, $y = 4 - (-5) = 9$

When $x = 2$, $y = 4 - 2 = 2$

Hence coordinates of points of intersection P and Q are $(-5, 9)$ and $(2, 2)$.

Using the discriminant to identify or show whether a line and circle intersect and, if so, how many times

If the circle and line do not meet or intersect, the resulting quadratic equation, when the two equations are solved simultaneously, will have no real roots.

To prove that there are no real roots of a quadratic equation in the form $ax^2 + bx + c = 0$ we can show that the discriminant $b^2 - 4ac < 0$.

Note also that:

If $b^2 - 4ac > 0$ there are two real and distinct roots, meaning the circle and line intersect in two places.

If $b^2 - 4ac = 0$ there are two real and equal roots (i.e. only one solution), meaning the circle and line meet in one place. The line is therefore a tangent to the circle.

Condition for two circles to touch internally or externally

When two circles touch externally it means that one of the circles is outside the other and they touch at a single point. When circles touch internally, one of the circles is inside the other.

Circles touching externally at a point

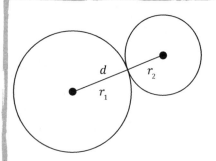

When two circles touch externally at a point, the distance between the centres of the circles must equal the sum of the radii of the two circles.

If the distance between the centres is d and the radii of the two circles are r_1 and r_2 then if the circles touch externally at a point:

$$d = r_1 + r_2$$

Example

Circle C_1 has centre $A(-2, 1)$ and radius 5. Circle C_2 has centre $B(10, 6)$ and radius r. If the circles C_1 and C_2 touch externally, find the value of r.

If d is the distance between the centres A and B, then

$d = \sqrt{(x_2 - x_1)^2 + (y_2 - y_1)^2}$

$d = \sqrt{(10 - (-2))^2 + (6 - 1)^2}$

$d = \sqrt{144 + 25}$

$d = \sqrt{169}$

$d = 13$

As the circles touch externally, the sum of the radii must equal the distance between the centres.

Hence, $13 = r + 5$, giving $r = 8$

Circles touching internally at a point

When two circles touch internally at a point, the distance between the centres of the circles equals the difference of the radii of the two circles.

If the distance between the centres is d and the radii of the two circles are r_1 and r_2 then if the circles touch internally at a point:

$$d = r_1 - r_2$$

Example

Two circles C_1 and C_2 touch internally at a point. Circle C_1 has the following equation:

$$x^2 + y^2 - 4x - 4y - 1 = 0$$

If circle C_2 has centre (2, 1) find the radius of this circle.

Answer

$$x^2 + y^2 - 4x - 4y - 1 = 0$$
$$(x-2)^2 + (y-2)^2 - 4 - 4 - 1 = 0$$
$$(x-2)^2 + (y-2)^2 = 9$$

Completing the square and rearranging to find the centre and radius of this circle. This equation is now in the form

$$(x-a)^2 + (y-b)^2 = r^2$$

A circle having the above equation will have centre (a, b) and radius r.

Alternatively you could use the equation of the circle in the form:

There is the following alternative form for the equation of a circle:

$$x^2 + y^2 + 2gx + 2fy + c = 0$$

A circle having the above equation will have centre (−g, −f) and radius given by

$$\sqrt{g^2 + f^2 - c}$$

Hence circle C_1 has centre (2, 2) and radius 3

Distance between the centres of both circles

$$d = \sqrt{(x_2 - x_1)^2 + (y_2 - y_1)^2}$$
$$= \sqrt{(2-2)^2 + (2-1)^2}$$
$$= 1$$

Now if the circles touch internally

$$d = r_1 - r_2$$
$$1 = 3 - r_2$$
$$r_2 = 2$$

Hence the radius of circle $C_2 = 2$

Example

① The circle C has centre A and equation

$x^2 + y^2 - 2x + 6y - 15 = 0$.

(a)(i) Write down the coordinates of A.

(ii) The point P has coordinates $(4, -7)$ and lies on C. Find the equation of the tangent to C at P. [5]

(b) The line L has equation $y = x + 4$. Show that L and C do not intersect. [4]

(WJEC C2 Jan 2011 Q8)

Answer

① (a)(i) $x^2 + y^2 - 2x + 6y - 15 = 0$.

Completing the square gives the following:

$(x - 1)^2 + (y + 3)^2 - 1 - 9 - 15 = 0$

$(x - 1)^2 + (y + 3)^2 = 25$

| You could alternatively use the formula to work out the coordinates of the centre and the radius of the circle. |

Comparing this with the equation of the circle $(x - a)^2 + (y - b)^2 = r^2$

gives the coordinates of centre A as $(1, -3)$ and radius $= 5$.

(ii) The line joining P and A will be a radius of the circle C.

Gradient of line joining $P(4, -7)$ and $A(1, -3) = \dfrac{-7 - (-3)}{4 - 1} = \dfrac{-4}{3}$

The radius AP will be the normal to the circle at P. The product of the gradients of the tangent and normal will be -1.

| A normal and a tangent make an angle of 90° to each other. |

Gradient of tangent at $P = \dfrac{3}{4}$ (i.e. using $m_1 m_2 = -1$)

Equation of the tangent at C having gradient $\dfrac{3}{4}$ and passing through the point $(4, -7)$ is

$y - (-7) = \dfrac{3}{4}(x - 4)$

$4y + 28 = 3x - 12$

$3x - 4y - 40 = 0$

| Remember: The equation of a straight line having gradient m and passing through the point (x_1, y_1) is given by: $y - y_1 = m(x - x_1)$ |

(b) Substituting $y = x + 4$ into the equation of the circle gives:

$x^2 + (x + 4)^2 - 2x + 6(x + 4) - 15 = 0$

Multiplying out the brackets and simplifying gives:

$2x^2 + 12x + 25 = 0$

Comparing this equation with $ax^2 + bx + c$, gives $a = 2$, $b = 12$ and $c = 25$.

Checking the roots of this equation:

$b^2 - 4ac = 144 - 4(2)(25) = 144 - 200 = -56$

As $b^2 - 4ac < 0$, there are no real roots which means the circle and line do not intersect.

| To find the points of intersection of a circle and a line you solve the two equations simultaneously. If there is one solution, then the line meets the circle at one point (i.e. it is a tangent). If there are two solutions, it cuts the circle in two places. If there are no solutions (because the resulting quadratic equation cannot be solved) the circle and line do not intersect or meet. |

Examination style questions

① The circle C has centre A and equation

$x^2 + y^2 - 4x + 6y = 3$

(a) Write down the coordinates of A and find the radius of C. [3]

(b) A straight line has equation $y = 4x - 7$.
This straight line intersects the circle C at two points. Find the coordinates of these two points. [4]

Answer

① (a) Comparing the equation $x^2 + y^2 - 4x + 6y = 3$ with the equation

$x^2 + y^2 + 2gx + 2fy + c = 0$ we can see $g = -2$, $f = 3$ and $c = -3$.

Centre A has coordinates $(-g, -f) = (2, -3)$

Radius $= \sqrt{g^2 + f^2 - c} = \sqrt{(-2)^2 + (3)^2 + 3} = \sqrt{16} = 4$

(b) Substituting $y = 4x - 7$ into the equation for the circle gives:

$x^2 + (4x - 7)^2 - 4x + 6(4x - 7) = 3$

$x^2 + 16x^2 - 56x + 49 - 4x + 24x - 42 = 3$

$17x^2 - 36x + 4 = 0$

$(17x - 2)(x - 2) = 0$

$x = \dfrac{17}{2}$ or $x = 2$

> The x-coordinates are substituted into the equation of the straight line to find the y-coordinates of the points of intersection.

When $x = \dfrac{17}{2}$, $y = 4\left(\dfrac{17}{2}\right) - 7 = 27$

When $x = 2$, $y = 4(2) - 7 = 1$

Hence the curve intersects the straight line at the points $\left(\dfrac{17}{2},\ 27\right)$ and $(2, 1)$.

② Circle C has centre A and radius r. The points $P(0, 5)$ and $Q(8, -1)$ are at either end of a diameter of C.

(a) (i) Write down the coordinates of A.

(ii) Show that $r = 5$.

(iii) Write down the equation of C.

[4]

(b) Verify that the point $R(7, 6)$ lies on C. [2]

(c) Find the equation of the tangent at point R. [3]

Answer

② (a) (i) A is the mid-point of PQ.

Hence coordinates of A are

$\left(\dfrac{0+8}{2},\ \dfrac{5+(-1)}{2}\right) = (4, 2)$

> The formula for the mid-point $\left(\dfrac{x_1 + x_2}{2},\ \dfrac{y_1 + y_2}{2}\right)$ is used here.

(ii)   The length of the straight line joining the two points $A(4, 2)$ and $P(0, 5)$ is given by:

Distance $AP = r = \sqrt{(x_2 - x_1)^2 + (y_2 - y_1)^2}$

$= \sqrt{(0-4)^2 + (5-2)^2}$

$= \sqrt{16 + 9}$

$= \sqrt{25}$

$= 5$

(iii)   Equation of the circle is

$(x-4)^2 + (y-2)^2 = 25$

> The equation of a circle having centre (a, b) and radius r is given by
> $$(x-a)^2 + (y-b)^2 = r^2$$

(b)   Substituting the coordinates of $R(7, 6)$ into the LHS of the equation gives:

LHS $= (7-4)^2 + (6-2)^2 = 9 + 16 = 25 = 5^2 =$ RHS, so the coordinates of R lie on the circle.

> Here you prove that the left-hand side of the equation, with the coordinates of the point entered for x and y, equals the right-hand side of the equation.

(c)   Gradient of line $AR = \dfrac{6-2}{7-4} = \dfrac{4}{3}$

Gradient of tangent $= -\dfrac{3}{4}$

> AR is a radius so it is perpendicular to the tangent at R.

Equation of tangent is

$y - 6 = -\dfrac{3}{4}(x-7)$

> The gradient of the tangent and a point through which it passes is substituted into the equation for a straight line.

$4y - 24 = -3x + 21$

$3x + 4y - 45 = 0$

Test yourself

Answer the following questions and check your answers before moving on to the next topic.

①   The circle C has equation $x^2 + y^2 - 8x - 6y = 0$.
The straight line L has equation $y + 2x + 4 = 0$.

(a)   Write down the coordinates of the centre of circle C and its radius.   [3]

(b)   Show that the straight line L and the circle C do not intersect or meet.   [4]

②   A circle has the equation $x^2 + y^2 - 4x + 6y = 3$.

(a)   Find the coordinates of the centre of the circle and its radius.

(b)   Show that the point $P(2, 1)$ lies on the circle.

③   Circle C has centre $A(2, 3)$ and radius 5.

(a)   Find the equation of circle C in the form

$x^2 + y^2 + ax + by + c = 0$

where a, b and c are constants to be determined.

(b)   Find the equation of the tangent to the circle at the point $P(5, 7)$.

(Note: answers to Test yourself are found at the back of the book.)

1   The circle C has centre A and equation $x^2 + y^2 - 8x + 2y + 7 = 0$

(a)   Find the coordinates of A and the radius of C.   [3]

(b)   The point P has coordinates $(7, -2)$.

(i)   Verify that P lies on C.

(ii)   Given that the point Q is such that PQ is a diameter of C, find the coordinates of Q.   [4]

(c)   The line L has equation $y = 2x - 4$. Find the coordinates of the points of intersection of L and C.   [4]

(WJEC C2 May 2010 Q9)

Answer

1 (a) $x^2 + y^2 - 8x + 2y + 7 = 0$

Completing the square gives $(x - 4)^2 + (y + 1)^2 - 16 - 1 + 7 = 0$

This gives $(x - 4)^2 + (y + 1)^2 = 10$

Comparing this with $(x - a)^2 + (y + b)^2 = r^2$ gives centre $A(4, -1)$ and radius $= \sqrt{10}$.

(b)   (i)   If point P lies on the circle, its coordinates will satisfy the equation of the circle.

Substituting the coordinates $(7, -2)$ into the LHS of the equation:

LHS $= 7^2 + (-2)^2 - 8(7) + 2(-2) + 7 = 49 + 4 - 56 - 4 + 7 = 0 = $ RHS

The coordinates of P satisfy the equation, proving that P lies on the circle.

(ii) $A(4, -1)$ and $P(7, -2)$

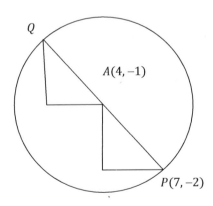

An alternative method of verifying that P lies on C involves showing that the distance of P from the centre A is the same as the radius of the circle, as follows:

$AP^2 = (x_1 - x_2)^2 + (y_1 - y_2)^2$

$\quad = (7 - 4)^2 + (-2 - (-1))^2$

$\quad = 3^2 + (-1)^2 = 9 + 1 = 10 = r^2 \Rightarrow AP = r$

The distance from the centre of the circle A to the point P is equal to the radius of C, therefore P lies on C.

As $AP = AQ$ to go from P to A the x-coordinate decreases by 3 and the y-coordinate increases by 1. We can use this fact when going from A to Q. Decreasing the x-coordinate of A by 3 gives 1 and increasing the y-coordinate by 1 gives 0. Hence the coordinates of Q are $(1, 0)$.

(c) $y = 2x - 4$ is substituted into the equation of the circle.

$x^2 + y^2 - 8x + 2y + 7 = 0$

$x^2 + (2x - 4)^2 - 8x + 2(2x - 4) + 7 = 0$

$x^2 + 4x^2 - 16x + 16 - 8x + 4x - 8 + 7 = 0$

$5x^2 - 20x + 15 = 0$

$x^2 - 4x + 3 = 0$

| Divide through by 5 to simplify this quadratic equation. |

$(x - 3)(x - 1) = 0$

$x = 3$ or $x = 1$

When $x = 3$, $y = 2(3) - 4 = 2$

When $x = 1$, $y = 2(1) - 4 = -2$

| Each x-coordinate is substituted into the equation of the straight line to find the corresponding y-coordinate. |

Hence, points of intersection are $(3, 2)$ and $(1, -2)$

Q&A 2

2   The circle C has centre A and radius r. The points $P(1, -4)$ and $Q(9, 10)$ are at either end of a diameter of C.

(a)   (i)   Write down the coordinates of A.

(ii) Show that $r = \sqrt{65}$.

(iii) Write down the equation of C.   [4]

(b)   Verify that the point $R(4, 11)$ lies on C.   [2]

(c)   Find $Q\hat{P}R$ [3]

(C2 May 2008 Q8)

Answer

2 (a)   (i)   Centre of circle is at the mid-point of the diameter PQ.

Mid-point of line joining $P(1, -4)$ and $Q(9, 10)$ is

$$\left(\frac{1+9}{2}, \frac{-4+10}{2} \right) = (5, 3)$$

(ii) Distance between the points $(1, -4)$ and $(5, 3)$ is given by

$r = \sqrt{(x_2 - x_1)^2 + (y_2 - y_1)^2}$

| The distance from the mid-point to the circumference is the radius of the circle. |

$r = \sqrt{(5 - 1)^2 + (3 - (-4))^2}$

$r = \sqrt{4^2 + 7^2}$

$= \sqrt{16 + 49}$

$r = \sqrt{65}$

(iii) The equation of a circle having centre (a, b) and radius r is given by

$(x - a)^2 + (y - b)^2 = r^2$

For this circle, centre is $(5, 3)$ and radius is $\sqrt{65}$.

$(x - 5)^2 + (y - 3)^2 = 65$

(b)   Substituting the coordinates of R into the LHS of the equation gives:

LHS $= (4-5)^2 + (11-3)^2 = (-1)^2 + (8)^2 = 65 =$ RHS, so the coordinates of R lie on the circle.

(c)

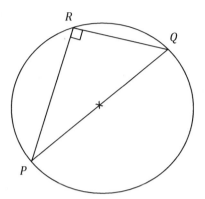

PQ is a diameter and as R is a point on the circumference, angle $P\hat{R}Q = 90°$ because it is an angle in a semicircle.

Length of $PQ = 2r = 2\sqrt{65}$

Length of $QR = \sqrt{(x_2 - x_1)^2 + (y_2 - y_1)^2}$

> The formula for the distance between two points is used here.

$\qquad\qquad = \sqrt{(9-4)^2 + (10-11)^2}$

$\qquad\qquad = \sqrt{26}$

Using trigonometry $\sin Q\hat{P}R = \dfrac{QR}{PQ}$

$\qquad\qquad\qquad = \dfrac{\sqrt{26}}{2\sqrt{65}}$

$\qquad\qquad\qquad = 0.3162$

Hence $Q\hat{P}R = \sin^{-1}(0.3162)$

$\qquad\qquad\quad = 18.4°$

Topic 4 — Trigonometry

This topic covers the following:

- The sine and cosine rules and the area of a triangle in the form $\dfrac{1}{2}ab\sin C$
- Radian measure, arc length, area of sector and area of segment
- Sine, cosine, and tangent functions, their graphs and periodicity
- Knowledge and use of $\tan\theta = \dfrac{\sin\theta}{\cos\theta}$ and $\cos^2\theta + \sin^2\theta = 1$
- Solution of simple trigonometric equations in a given interval

Sine, cosine and tangent functions and their exact values

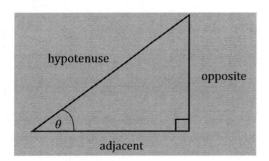

The following ratios, covered in your GCSE studies, only apply to right-angled triangles:

$$\sin\theta = \frac{\text{opposite}}{\text{hypotenuse}}$$

$$\cos\theta = \frac{\text{adjacent}}{\text{hypotenuse}}$$

$$\tan\theta = \frac{\text{opposite}}{\text{adjacent}}$$

The exact values of the sine, cosine and tangent of 30°, 45° and 60°

The exact values of the above angles can be determined by drawing triangles, working out the lengths of the sides that aren't known and then using trigonometry to work out the exact values of the angles.

The exact values of the sine, cosine and tangent of 45°

The exact values can be worked out by drawing the following triangle:

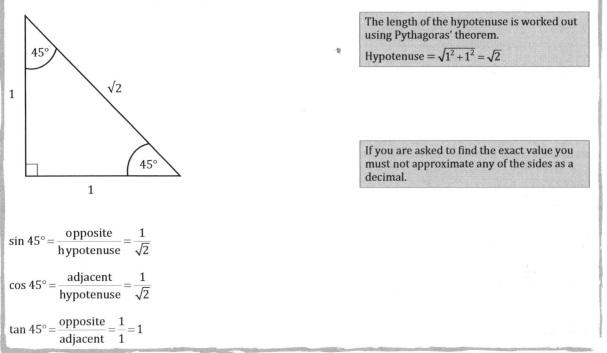

The length of the hypotenuse is worked out using Pythagoras' theorem.

$$\text{Hypotenuse} = \sqrt{1^2 + 1^2} = \sqrt{2}$$

If you are asked to find the exact value you must not approximate any of the sides as a decimal.

$$\sin 45° = \frac{\text{opposite}}{\text{hypotenuse}} = \frac{1}{\sqrt{2}}$$

$$\cos 45° = \frac{\text{adjacent}}{\text{hypotenuse}} = \frac{1}{\sqrt{2}}$$

$$\tan 45° = \frac{\text{opposite}}{\text{adjacent}} = \frac{1}{1} = 1$$

The exact values of the sine, cosine and tangent of 30° and 60°

The exact values can be worked out using an equilateral triangle having sides of length 2 and then drawing in one of the lines of symmetry to form two identical right-angled triangles:

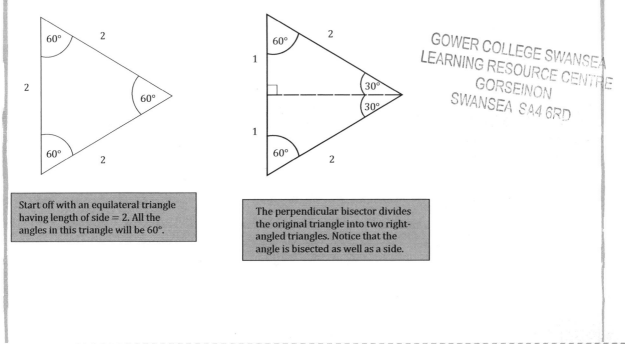

Start off with an equilateral triangle having length of side = 2. All the angles in this triangle will be 60°.

The perpendicular bisector divides the original triangle into two right-angled triangles. Notice that the angle is bisected as well as a side.

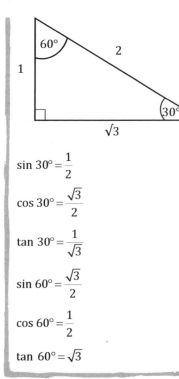

The ratio of lengths of sides is then used to give the sine, cosine and tangent of the various angles.

Half of the original triangle is used. The length of the base of this triangle is worked out using Pythagoras' theorem.

$$\text{Base} = \sqrt{2^2 - 1^2} = \sqrt{3}$$

$\sin 30° = \dfrac{1}{2}$

$\cos 30° = \dfrac{\sqrt{3}}{2}$

$\tan 30° = \dfrac{1}{\sqrt{3}}$

$\sin 60° = \dfrac{\sqrt{3}}{2}$

$\cos 60° = \dfrac{1}{2}$

$\tan 60° = \sqrt{3}$

Obtaining angles given a trigonometric ratio

Finding angles using the CAST method

One method of obtaining angles given a trigonometric ratio is called the CAST method.

The CAST method uses the diagram shown below. A indicates that all the ratios are positive in the first quadrant (for angles between 0° and 90°), S indicates that sine is positive in the second quadrant (90° to 180°), T indicates that tangent is positive in the third quadrant (180° to 270°) and C indicates that cosine is positive in the fourth quadrant (270° to 360°).

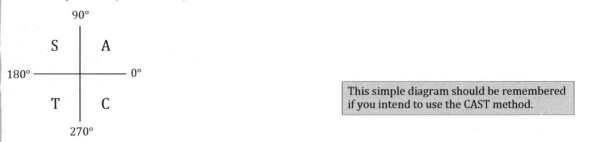

This simple diagram should be remembered if you intend to use the CAST method.

CAST stands for Cos, All, Sin, Tan and the diagram shows where these functions are positive. For example suppose we wanted all the values of the angle θ in the range $0° \le \theta \le 360°$ where $\sin \theta = 0.6946$. Here we have a positive value for $\sin \theta$. The sine function is positive in the first and second quadrants.

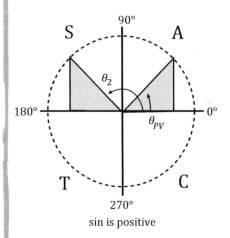

sin is positive

Notice that the angles are measured anticlockwise from 0°.

Two triangles are drawn in the regions where $\sin \theta$ is positive. You can use your calculator to find the first value θ_{PV} by working out $\sin^{-1}(0.6946)$ using a calculator. This gives $\theta_{PV} = 44°$. As both triangles are identical the value of θ_2 is found by subtracting 44° from 180°. Hence the other angle is $180° - 44° = 136°$. Therefore $\theta = 44°$ or $136°$.

Example

① Find all the values of the angle θ in the range $0° \leq \theta \leq 360°$ where $\cos \theta = -\dfrac{1}{2}$.

Answer

① Two triangles are drawn in the regions where $\cos \theta$ is negative. Cosine is negative in the second and third quadrants. θ_{PV} can be found by using a calculator and entering $\cos^{-1}(-0.5)$ giving a value of 120°. By symmetry, the value of θ_2 can be found by subtracting 120° from 360°. Hence the other angle is $360° - 120° = 240°$. Therefore $\theta = 120°$ or $240°$.

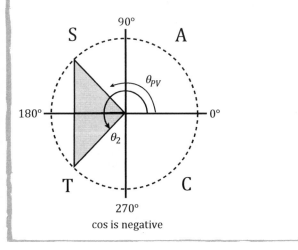

cos is negative

Using this method, the solutions are found by making the same angle from the horizontal in each of the appropriate quadrants, e.g. $180° - 60° = 120°$ and $180° + 60° = 240°$

Finding angles using trigonometric graphs

Another method involves using the trigonometric graphs to find all the angles. Here you must be able to draw the graphs of each trigonometric function (sin, cos and tan).

You will probably be familiar with these graphs from your GCSE work. The graphs are included on pages 135 to 136.

Example

② Find all the values of the angle θ in the range $0° \le \theta \le 360°$ where $\sin \theta = \dfrac{1}{2}$.

Answer

② The graph of $y = \sin\theta$ is drawn in the range $0° \le \theta \le 360°$.

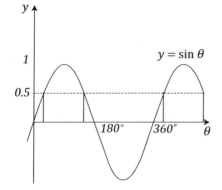

The first angle is found by performing the calculation $\sin^{-1}\left(\dfrac{1}{2}\right)$ or $\sin^{-1}(0.5)$ using a calculator or by recognition (see page 125). The result is 30°. By the symmetry of the graph you can see that the other angle will be $180° - 30° = 150°$. Hence the two values of θ in the required range are 30° and 150°.

The sine and cosine rules

The sine and cosine rules can be used with any triangle, not just those containing a right angle.

The angles are denoted by the letters A, B and C and the lengths of the sides opposite these angles are denoted by a, b and c respectively.

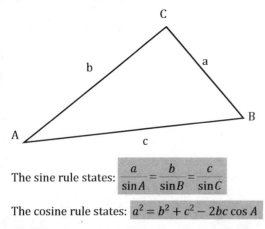

> The sine rule should be used in the most useful form for the given question:
> $$\frac{a}{\sin A} = \frac{b}{\sin B} = \frac{c}{\sin C} \quad \text{or} \quad \frac{\sin A}{a} = \frac{\sin B}{b} = \frac{\sin C}{c}$$

The sine rule states: $\dfrac{a}{\sin A} = \dfrac{b}{\sin B} = \dfrac{c}{\sin C}$

The cosine rule states: $a^2 = b^2 + c^2 - 2bc \cos A$

> The formulae for the sine and cosine rules are not included in the formula booklet so must be memorised.

The area of a triangle

If two sides of a triangle are known as well as the included angle, then the area of the triangle can be found using the formula:

Area of triangle $= \dfrac{1}{2}\, ab \sin C$

> This formula works for all triangles but for areas of right-angled triangles, use the formula $A = \dfrac{1}{2} \times base \times height$
> Note that you will not be given this formula.

Warning: be careful when using this formula to work out the angle when the area and two sides of the triangle are known. For example $\sin C = \dfrac{1}{2}$ can have two solutions 30° and 150°. If another angle in the triangle is known then the obtuse angle may not be a possible solution. Be guided by the wording of the question, so look for the plural 'angles' in the question to see if you are looking for two possible angles.

Example

① The diagram below shows the triangle ABC with $AB = x$ cm, $AC = (x + 4)$ cm and $B\hat{A}C = 150°$.

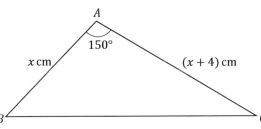

Given that the area of the triangle ABC is 15 cm^2,

(a) find the value of x, [3]

(b) find the length of BC correct to one decimal place. [2]

(WJEC C2 May 2008 Q3)

Answer

① (a) Area of triangle $ABC = \dfrac{1}{2}\, bc\, \sin A$

$= \dfrac{1}{2}(x+4)x \sin 150°$

$= \dfrac{1}{2}(x+4)x\left(\dfrac{1}{2}\right)$

$= \dfrac{1}{4}\left(x^2 + 4x\right)$

$15 = \dfrac{1}{4}\left(x^2 + 4x\right)$

$60 = x^2 + 4x$

$x^2 + 4x - 60 = 0$

$(x - 6)(x + 10) = 0$

Solving gives $x = 6$ or $x = -10$

As x is a length of a side, $x = -10$ is impossible, hence $x = 6$

(b) Substituting the value $x = 6$ for the two sides gives

$AB = 6$ cm and $AC = 10$ cm

Using the cosine rule

$BC^2 = 6^2 + 10^2 - 2 \times 6 \times 10 \cos 150°$

$BC^2 = 36 + 100 - 120 \cos 150°$

$BC^2 = 136 + 103.92$

$BC = \sqrt{239.92}$

$BC = 15.5$ cm (correct to 1 decimal place)

Example

② The triangle ABC is such that $AB = 16$ cm, $AC = 9$ cm and $A\hat{B}C = 23°$.

(a) Find the possible values of $A\hat{C}B$. Give your answers correct to the nearest degree. [2]

(b) Given that $B\hat{A}C$ is an **acute** angle, find:

    (i) the size of $B\hat{A}C$, giving your answer correct to the nearest degree,

    (ii) the area of triangle ABC, giving your answer correct to one decimal place. [4]

(WJEC C2 May 2009 Q3)

Answer

② (a)

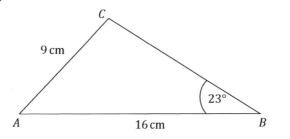

It is a good idea to draw a sketch showing what the triangle might look like. It will enable you to see which sides and angles are known so you can decide whether to use the sine or cosine rule.

Using the sine rule

$\dfrac{\sin A\hat{C}B}{16} = \dfrac{\sin 23°}{9}$

$\sin A\hat{C}B = \dfrac{16 \sin 23°}{9}$

$\sin A\hat{C}B = 0.6946$

$A\hat{C}B = 44°$ or $136°$

The question does not give you a diagram for the triangle and notice that it asks for possible **values** of $A\hat{C}B$. This implies more than one value, so you have to look for the two angles that have a sine of 0.6946.

(b) (i) If angle BAC is acute, angle ACB must be $136°$

Hence, angle $BAC = 180° - (136° + 23°) = 21°$

If angle ACB was $44°$ it would mean that angle BAC would be obtuse.

(ii) Area of triangle $= \dfrac{1}{2} bc \sin A$

$\qquad = \dfrac{1}{2} \times 9 \times 16 \times \sin 21°$

$\qquad = 25.8025 \text{ cm}^2$

$\qquad = 25.8 \text{ cm}^2$ (to one decimal place)

Example

③ The diagram below shows a sketch of the triangle ABC with $AB = 8$ cm, $AC = x$ cm, $BC = (x + 2)$ cm and $A\hat{B}C = 60°$.

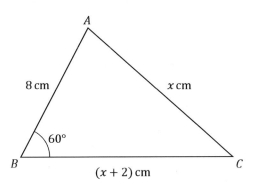

(a) Write down and simplify an equation satisfied by x. Hence evaluate x.   [3]

(b) Find the size of $A\hat{C}B$.   [2]

(WJEC C2 Jan 2010 Q3)

Answer

③ (a) Using the cosine rule

$\qquad x^2 = 8^2 + (x + 2)^2 - 2 \times 8 \times (x + 2) \cos 60°$

$\qquad x^2 = 64 + x^2 + 4x + 4 - 16 \times (x + 2) \times \dfrac{1}{2}$

$\qquad x^2 = 64 + x^2 + 4x + 4 - 8x - 16$

$\qquad x^2 = x^2 - 4x + 52$

$\qquad 0 = -4x + 52$

$\qquad x = 13$

(b)

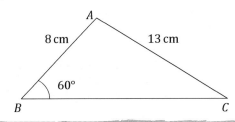

Using the sine rule

$$\frac{\sin C}{c} = \frac{\sin B}{b}$$

$$\frac{\sin A\hat{C}B}{8} = \frac{\sin 60°}{13}$$

$$\sin A\hat{C}B = \frac{8 \times \sin 60°}{13} = 0.5329$$

Hence $A\hat{C}B = 32.2°$

Radian measure, arc length, area of sector and area of segment

Radian measure

There is another unit for measuring angles called the radian.

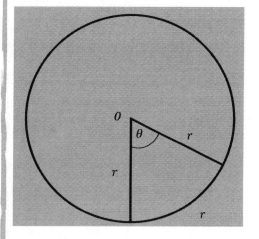

When the length of an arc is the same as the radius then the angle between the two radii and the arc, θ, is 1 radian. An arc of length $2r$ would give an angle at the centre of 2 radians and an arc of length θr would give an angle at the centre of θ radians.

If the length of the arc equals half the circumference then the length of the arc is πr. If this length of arc corresponds to an angle at the centre of θ radians then equating the two gives:

Equating the two gives:

$r\theta = \pi r$

So $\theta = \pi$ (now as $\theta = 180°$) we can write π radians $= 180°$

So 1 radian $= \dfrac{180}{\pi} = \dfrac{180}{3.14} = 57.3°$

Here are some popular angles expressed in radians and degrees:

2π radians $= 360°$

$\dfrac{\pi}{2}$ radians $= 90°$ $\dfrac{\pi}{4}$ radians $= 45°$

$\dfrac{\pi}{3}$ radians $= 60°$ $\dfrac{\pi}{6}$ radians $= 30°$

> Check that your calculator is set to radian mode when performing radian calculations. Remember to turn it back to degrees after completing the question.

Arc length

The length of an arc making an angle of θ radians at the centre $l = r\theta$

l is a fraction of the circumference and is given by

$$l = \frac{\theta}{2\pi} \times 2\pi r = r\theta$$

> Remember that for arc lengths and areas of sectors the angle at the centre is measured in radians.

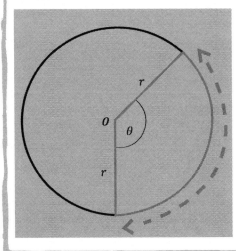

Area of a sector

Area of a sector making an angle of θ radians at the centre $= \frac{1}{2}r^2\theta$

A is a fraction of the area of the complete circle and is given by

$$A = \frac{\theta}{2\pi} \times \pi r^2 = \frac{1}{2}r^2\theta$$

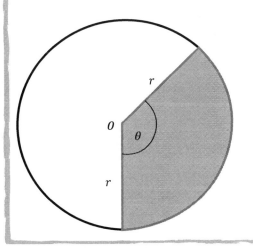

Area of a segment

A segment of a circle is the area bounded by a chord and an arc. It is the shaded area in this diagram.

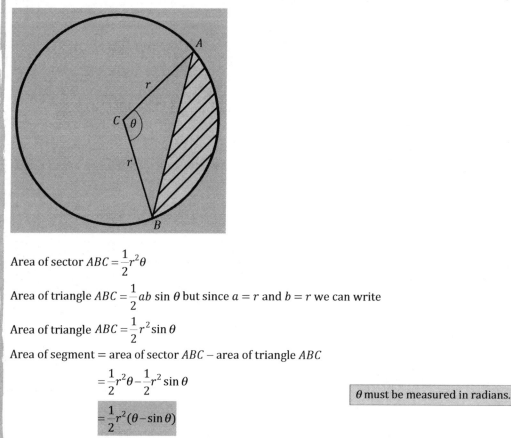

Area of sector $ABC = \dfrac{1}{2}r^2\theta$

Area of triangle $ABC = \dfrac{1}{2}ab\sin\theta$ but since $a = r$ and $b = r$ we can write

Area of triangle $ABC = \dfrac{1}{2}r^2\sin\theta$

Area of segment = area of sector ABC − area of triangle ABC

$$= \frac{1}{2}r^2\theta - \frac{1}{2}r^2\sin\theta$$

$$= \frac{1}{2}r^2(\theta - \sin\theta)$$

θ must be measured in radians.

Example

①

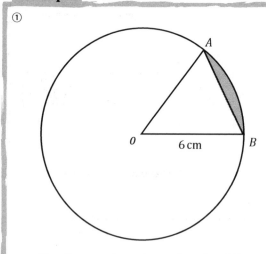

The diagram shows two points A and B on a circle with centre O and radius 6 cm. The length of the **arc** AB is 5.4 cm.

(a) Show that the area of the **sector** AOB is 16.2 cm^2.   [4]

(b) Find the area of the shaded region, giving your answer correct to one
decimal place.   [3]

(WJEC C2 May 2008 Q9)

Answer

① (a) Length of arc $AB = r\theta$

$5.4 = 6\theta$

$\theta = 0.9$ radians

Area of sector $AOB = \dfrac{1}{2}r^2\theta$

$= \dfrac{1}{2} \times 6^2 \times 0.9$

$= 16.2$ cm^2

(b) Area of triangle $AOB = \dfrac{1}{2}ab\sin C$

$= \dfrac{1}{2} \times 6 \times 6 \sin 0.9$

$= 14.10$ cm^2

Area of shaded region (i.e. the segment) = Area of sector − Area of triangle

$= 16.2 - 14.10$

$= 2.1$ cm^2 (correct to 1 decimal place)

> The angle at the centre is not given. The length of arc formula can be used to work out the angle at the centre in radians.

> ### ⤊ Grade boost
> Check the formulae in the formula booklet regularly so you know which you need to remember. Note that the formulae for the length of an arc and the area of a sector are not given in the formula booklet.

> Note that both a and b are equal to the radius r of the circle.

> Remember to set your calculator to radian mode before working out this calculation.

Sine, cosine and tangent graphs and their periodicity

The sine graph ($y = \sin\theta$) where θ is expressed in radians

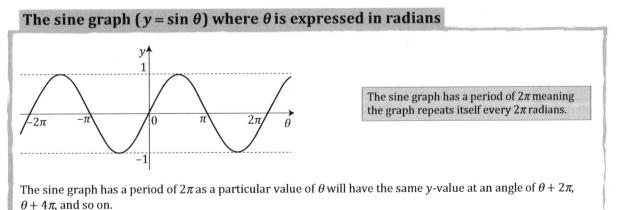

> The sine graph has a period of 2π meaning the graph repeats itself every 2π radians.

The sine graph has a period of 2π as a particular value of θ will have the same y-value at an angle of $\theta + 2\pi$, $\theta + 4\pi$, and so on.

The sine graph ($y = \sin \theta$) where θ is expressed in degrees

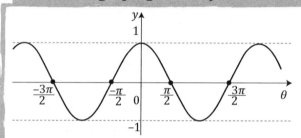

The cosine graph ($y = \cos \theta$) where θ is expressed in radians

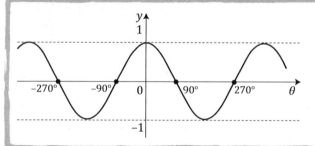

The cosine graph has a period of 2π meaning the graph repeats itself every 2π radians.

The cosine graph has a period of 2π as a particular value of θ will have the same y-value at an angle of $\theta + 2\pi$, $\theta + 4\pi$, and so on.

The cosine graph ($y = \cos \theta$) where θ is expressed in degrees

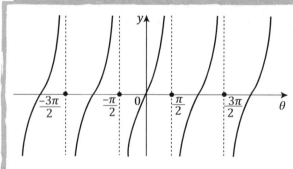

The tangent graph ($y = \tan \theta$) where θ is expressed in radians

The period of the tan θ graph is π radians.

The tangent graph ($y = \tan \theta$) where θ is expressed in degrees

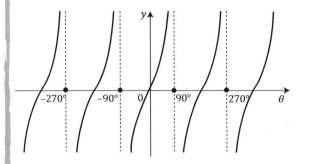

Knowledge and use of $\tan \theta = \dfrac{\sin \theta}{\cos \theta}$ and $\cos^2 \theta + \sin^2 \theta = 1$

There are two trigonometric identities that you may need to use when solving simple trigonometric equations. These two identities are:

$\tan \theta = \dfrac{\sin \theta}{\cos \theta}$

$\cos^2 \theta + \sin^2 \theta = 1$

You will see both of the above identities being used in the examples following the next section.

> Both of these identities must be remembered. They are not included in the formula booklet.

> You may see the identity symbol $\equiv$ being used sometimes instead of the equals sign. You can treat them as being the same. An identity is a relationship involving a letter which is true for all values of the letter.

Solution of simple trigonometric equations in a given interval

The graphs of trigonometric functions can be used to help identify all the solutions of a simple trigonometric equation in a given interval.

Suppose we have to solve the following equation in the interval $0° \leq \theta \leq 360°$

$\sin(2x - 30)° = \dfrac{1}{2}$

Draw a graph of $y = \sin \theta$. You will need to go a lot further than 360° for your graph so that all the possible solutions are shown.

Here we will go as far as 720°

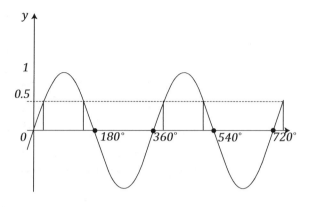

$$2x - 30 = \sin^{-1}\left(\frac{1}{2}\right)$$

Letting $\theta = 2x - 30$

So $\theta = \sin^{-1}\left(\frac{1}{2}\right)$

> You could also use the CAST method for finding all the angles θ in the required range.

Using the calculator (or the triangles for the exact values learnt earlier) we have

$\theta = 30°$ we can then see from the graph that the line $y = \dfrac{1}{2}$ also intersects the curve at the following values of θ:
$\theta = 150°, 390°, 510°$

When looking for all the solutions, between 0° and 360°, of $(2\theta - 30) = \dfrac{1}{2}$, the values of θ that need to be considered will lie between $2 \times 0 - 30$ and $2 \times 360 - 30$, i.e. −30° and 690°.

In practice, it's probably best to consider twice the range, since the multiple of the angle involved is 2, i.e. from 0° to 720°. It is quicker to do this calculation.

Hence $\theta = 30°, 150°, 390°, 510°$

So $2x - 30 = 30°, 150°, 390°, 510°$

> Sine is positive in the first and second quadrants, therefore $\theta = 30°$ or $\theta = 180 - 30 = 150°$, or (adding 360°) 390° or 510°, etc.

$2x = 60°, 180°, 420°, 540°$

$x = 30°, 90°, 210°, 270°$

So values of x in the required range are $x = 30°, 90°, 210°, 270°$

Example

① Find the values of x in the range $0° \leq \theta \leq 360°$, that satisfy the equation

   $2 \sin x = \tan x$

Answer

① $2 \sin x = \tan x$

   $2\sin x = \dfrac{\sin x}{\cos x}$

   $2 \sin x \cos x = \sin x$

> Do not be tempted here to divide both sides by sin x. If you do this then you will lose some solutions of the equation.

$2 \sin x \cos x - \sin x = 0$

$\sin x (2 \cos x - 1) = 0$

Hence, $\sin x = 0$ or $2 \cos x - 1 = 0$

So $\sin x = 0$ or $\cos x = \dfrac{1}{2}$

$\sin x = 0$ at $x = 0°, 180°, 360°$

$\cos x = \dfrac{1}{2}$ at $x = 60°$ or $300°$

Hence $x = 0°, 60°, 180°, 300°$ or $360°$

> Using the CAST method e.g. for $\cos x = \dfrac{1}{2}$, cosine is positive in the first and fourth quadrants, so $x = 60°$ or $x = 360° - 60° = 300°$.

Example

② (a) Find all values of θ in the range $0° \leq \theta \leq 360°$ satisfying $12 \cos^2 \theta - 5 \sin \theta = 10$ [6]

(b) Find all values of x in the range $0° \leq x \leq 180°$ satisfying $\tan 2x = -1.6$ [2]

(c) Find all values of ϕ in the range $0° \leq \phi \leq 180°$ satisfying $\tan \phi + 2 \sin \phi = 0$ [4]

(WJEC C2 May 2010 Q2)

Answer

② (a) $12 \cos^2 \theta - 5 \sin \theta = 10$

$12(1 - \sin^2 \theta) - 5 \sin \theta = 10$

$12 \sin^2 \theta + 5 \sin \theta - 2 = 0$

$(4 \sin \theta - 1)(3 \sin \theta + 2) = 0$

$\sin \theta = \dfrac{1}{4}$ or $\sin \theta = -\dfrac{2}{3}$

When $\sin \theta = \dfrac{1}{4}$, $\theta = 14.5°$ or $165.5°$

When $\sin \theta = -\dfrac{2}{3}$, $\theta = 221.8°$ or $318.2°$

$\theta = 14.5°, 165.5°, 221.8°$ or $318.2°$

> $\cos^2 \theta + \sin^2 \theta = 1$ so $\cos^2 \theta = 1 - \sin^2 \theta$

> Remember to include all the solutions in the required range.

(b) $\tan 2x = -1.6$

$2x = 122°, 302°$

$x = 61°$ or $151°$

(c) $\tan \phi + 2 \sin \phi = 0$

$\dfrac{\sin \phi}{\cos \phi} + 2 \sin \phi = 0$

$\sin \phi + 2 \sin \phi \cos \phi = 0$

$\sin \phi (1 + 2 \cos \phi) = 0$

$\sin \phi = 0$ or $\cos \phi = -\dfrac{1}{2}$

$\phi = 0°, 180°$ or $120°$

> Use $\tan \phi = \dfrac{\sin \phi}{\cos \phi}$

Examination style questions

①   (a)   Find all the values of θ in the range $0° \leq \theta \leq 360°$ satisfying $2 \sin \theta = 1$ [3]

     (b)   Find all the values of θ in the range $0° \leq \theta \leq 2\pi$ satisfying $\tan \dfrac{\theta}{2} = \sqrt{3}$
giving your answers in terms of π.   [3]

Answer

①   (a) $2 \sin \theta = 1$

$$\sin \theta = \frac{1}{2}$$

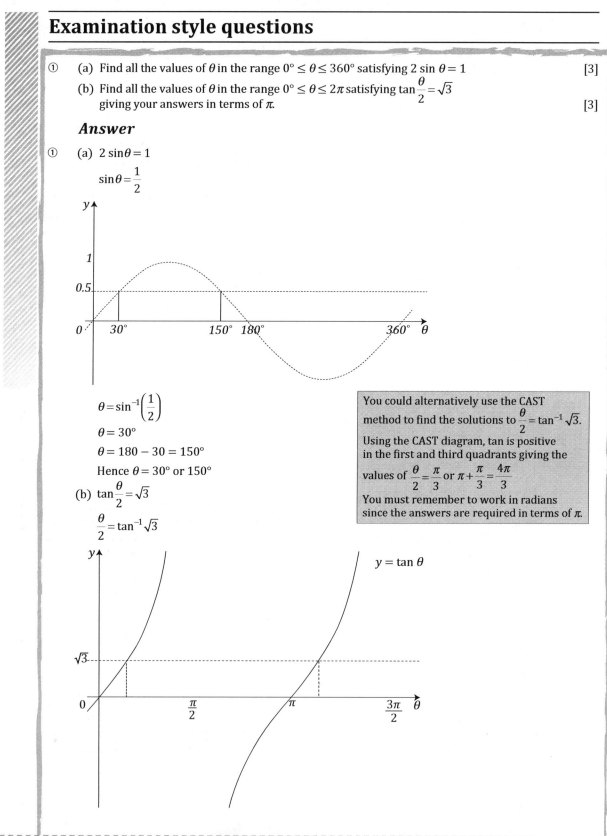

$$\theta = \sin^{-1}\left(\frac{1}{2}\right)$$

$$\theta = 30°$$

$$\theta = 180 - 30 = 150°$$

Hence $\theta = 30°$ or $150°$

     (b) $\tan \dfrac{\theta}{2} = \sqrt{3}$

$$\frac{\theta}{2} = \tan^{-1}\sqrt{3}$$

> You could alternatively use the CAST method to find the solutions to $\dfrac{\theta}{2} = \tan^{-1}\sqrt{3}$.
> Using the CAST diagram, tan is positive in the first and third quadrants giving the values of $\dfrac{\theta}{2} = \dfrac{\pi}{3}$ or $\pi + \dfrac{\pi}{3} = \dfrac{4\pi}{3}$
> You must remember to work in radians since the answers are required in terms of π.

$y = \tan \theta$

$$\frac{\theta}{2} = \tan^{-1}\sqrt{3}$$

Since values of θ are asked for in the range

$0 \le \theta \le 2\pi$, then $\dfrac{\theta}{2}$ will be in the range 0 to π, so

there is no need to consider values greater than π

Hence $\dfrac{\theta}{2} = \dfrac{\pi}{3}$

$\theta = \dfrac{2\pi}{3}$ radians

> $\text{Tan}\dfrac{\pi}{3} = \sqrt{3}$ this result should be known.
> Remember that $\dfrac{\pi}{3} = 60°$.

②

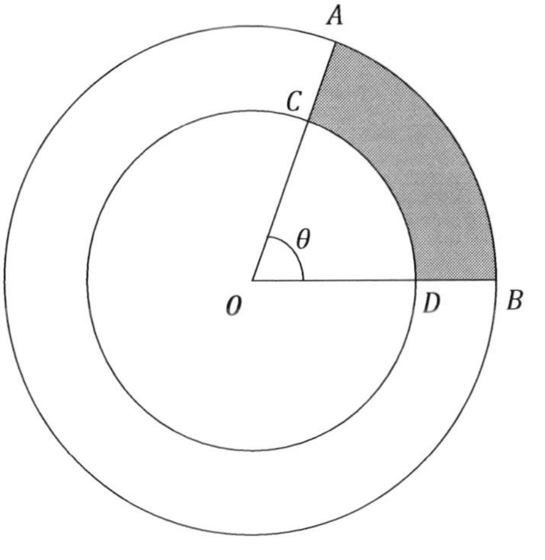

The diagram shows two concentric circles with common centre O. The radius of the larger circle is R cm and the radius of the smaller circle is r cm. The points A and B lie on the larger circle and are such that $A\hat{O}B = \theta$ radians. The smaller circle cuts OA and OB at the points C and D respectively. The length of the arc AB is L cm **greater** than the length of the arc CD. The area of the shaded region is K cm^2.

    (a) (i)   Write down an expression for L in terms of R, r and θ.

       (ii) Write down an expression for K in terms of R, r and θ.   [2]

    (b) Use your results to part (a) to find an expression for r in terms of R, K and L.   [3]

    (WJEC C2 May 2010 Q10)

Answer

②   (a) (i)   Arc AB − Arc $CD = L$

             $R\theta - r\theta = L$

           Hence $L = \theta(R - r)$

> Remember that the length of an arc $= r\theta$ where r is the radius of the circle and θ is the angle at the centre measured in radians. This formula needs to be remembered.

(ii)   Shaded area = Area of sector OAB − Area of sector OCD

$$K = \frac{1}{2}R^2\theta - \frac{1}{2}r^2\theta$$

$$K = \frac{1}{2}\theta(R^2 - r^2)$$

(b) $K = \frac{1}{2}\theta(R^2 - r^2) = \frac{1}{2}\theta(R - r)(R + r)$

> $R^2 - r^2$ is the difference of two squares and can be factorised.

Now $L = \theta(R - r)$

Rearranging gives $\theta = \dfrac{L}{R - r}$

Substituting this into the equation for K gives

$$K = \frac{1}{2}\frac{L}{(R - r)}(R - r)(R + r)$$

Cancelling $(R - r)$ on the top and bottom gives

$$K = \frac{1}{2}L(R + r)$$

> When cancelling $(R - r)$ it is valid because $R - r \neq 0$ or $R \neq r$, as otherwise there would be no shaded region.

$$\frac{2K}{L} = R + r$$

Giving $r = \dfrac{2K}{L} - R$

Test yourself

Answer the following questions and check your answers before moving on to the next topic.

①   The diagram shows the triangle ABC with $AB = 8$ cm, $AC = 12$ cm and $B\hat{A}C = 150°$.

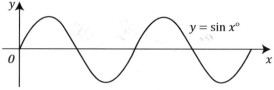

(a) Find the area of triangle ABC.

(b) Find the length of BC correct to one decimal place.

②   The graph shows the curve $y = \sin x$ in the interval $0 \leq x \leq 4\pi$.

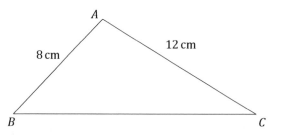

(a) Write down the coordinates of all the points of intersection with the x-axis.

(b) Write down the coordinates of all stationary points for this graph.

③   The triangle ABC is such that $AB = 4$ cm, $BC = \left(3\sqrt{2}-1\right)$ cm and $B\hat{A}C = 30°$.

Find an expression for sin $A\hat{C}B$ in the form $\dfrac{2+m\sqrt{2}}{n}$, where m, n are integers whose values are to be found.

④   Find all the values of θ in the range $0° \leq \theta \leq 360°$ satisfying

$3 \sin^2 \theta = 5 - 5 \cos \theta$

(Note: answers to Test yourself are found at the back of the book.)

1 (a) Find all values of θ between 0° and 360° satisfying $7\sin^2\theta + 1 = 3\cos^2\theta - \sin\theta$ [6]

(b) Find all values of x between 0° and 180° satisfying $\cos(2x + 25°) = -0.454$ [3]

(WJEC C2 Jan 2011 Q2)

Answer

1 (a) $7\sin^2\theta + 1 = 3\cos^2\theta - \sin\theta$

$7\sin^2\theta + 1 = 3(1 - \sin^2\theta) - \sin\theta$

$7\sin^2\theta + 1 = 3 - 3\sin^2\theta - \sin\theta$

$10\sin^2\theta + \sin\theta - 2 = 0$

$(5\sin\theta - 2)(2\sin\theta + 1) = 0$

$\sin\theta = \dfrac{2}{5}$ or $\sin\theta = -\dfrac{1}{2}$

When $\sin\theta = \dfrac{2}{5}$, $\theta = 23.6°, 156.4°$

When $\sin\theta = -\dfrac{1}{2}$, $\theta = 210°, 330°$

Hence $\theta = 23.6°, 156.4°, 210°$ or $330°$

> Although each of the terms in $\sin^2\theta$ or $\cos^2\theta$ could be written in terms of the other using $\sin^2\theta + \cos^2\theta = 1$, this is not as straightforward for the term in $\sin\theta$. Since there is a term in $\sin\theta$, we shall write the equation just in terms of $\sin\theta$.

> Using the CAST method Sin is positive in the first and second quadrants so $\theta = 23.6°$ or $180° - 23.6° = 156.4°$. Alternatively you could draw a sine graph to help work out the values.

> When $\sin\theta = -\dfrac{1}{2}$, sine is negative in the third and fourth quadrants, so $\theta = 180° + 30°$ or $360° - 30°$, $\theta = 210°$ or $330°$

(b) $\cos(2x + 25°) = -0.454$

$2x + 25° = 117°, 243°$

$2x = 92°, 218°$

$x = 46°, 109°$

Possible values of x are 46° and 109°

> Since the question asks for all the solutions between 0° and 180°, and the angle involved in the question is $2x$, then values of $2x$ need to be considered between 0° and 360° only.

2

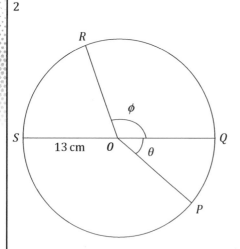

The diagram shows four points P, Q, R and S on a circle with centre O and radius 13 cm. The line QS is a diameter of the circle, $P\hat{O}Q = \theta$ radians and $Q\hat{O}R = \phi$ radians.

(a)   The area of sector POQ is 60 cm^2.

Find the value of θ, giving your answer correct to two decimal places.   [2]

(b)   The length of the arc QR is 7 cm greater than the length of the arc RS.

Find the value of ϕ, giving your answer correct to two decimal places.   [3]

(WJEC C2 May 2009 Q9)

Answer

2 (a)   Area of sector $POQ = \dfrac{1}{2}r^2\theta$

$60 = \dfrac{1}{2} \times 13^2\theta$

$60 = \dfrac{1}{2} \times 169\theta$

$\theta = \dfrac{120}{169} = 0.71$ radians (to 2 decimal places)

(b)   Length of arc $QR = 13\phi$

Length of arc $RS = r(\pi - \phi) = 13(\pi - \phi)$

$QR - RS = 7$

$13\phi - 13(\pi - \phi) = 7$

$13\phi - 40.841 + 13\phi = 7$

$26\phi = 47.841$

$\phi = \dfrac{47.841}{26} = 1.84$ radians (to 2 decimal places)

> Remember here that angles have to be in radians. Be careful to use π radians here rather than 180°.

145

Topic 5 — Integration

This topic covers the following:

- Indefinite integration as the reverse process of differentiation
- Integration of x^n ($n \neq -1$)
- Approximation of the area under a curve using the trapezium rule
- Interpretation of the definite integral as the area under a curve
- Evaluation of definite integrals

Indefinite integration as the reverse process of differentiation

Integration is the opposite of differentiation.

For example: If $y = x^2 + 3x + 5$, then $\dfrac{dy}{dx} = 2x + 3$, so $\int 2x + 3 \, dx = x^2 + 3x + c$. Notice why a constant of integration called c is needed. When differentiating any constant terms disappear so when integrating it is impossible to know what the value of the constant term should be. Hence a constant c is added. Later you will see how the value of this constant can be found in certain circumstances.

So to integrate x^n you increase the index by one and then divide by the new index. It is important to note that this works for all values of n provided $n \neq -1$. For indefinite integration you must always remember to include the constant of integration, called c.

This can be expressed in the following way:

$$\int x^n dx = \frac{x^{n+1}}{n+1} + c \text{ (provided } n \neq -1)$$

You will see how this works by looking at the following examples:

1. $\int x^3 dx = \dfrac{x^4}{4} + c$

2. $\int 2x \, dx = \dfrac{2x^2}{2} + c = x^2 + c$

3. $\int 4 dx = 4x + c$

The following expression is integrated as follows:

$$\int (x^3 + 4x^2 - x + 2) dx = \frac{x^4}{4} + \frac{4x^3}{3} - \frac{x^2}{2} + 2x + c$$

> This is called an indefinite integral because the result is not definite and you must remember to add a constant of integration c.

Fractional powers can be integrated in the following way

$$\int \left(x^{\frac{1}{2}} + x^{\frac{1}{3}} \right) dx = \frac{x^{\frac{3}{2}}}{\frac{3}{2}} + \frac{x^{\frac{4}{3}}}{\frac{4}{3}} + c = \frac{2}{3}x^{\frac{3}{2}} + \frac{3}{4}x^{\frac{4}{3}} + c$$

> Remember that when you divide by a fraction you must invert the fraction and then multiply to give the answer.

If you have an integral with roots or reciprocals, you must change them to indices before integrating as the following example shows

$$\int \left(\sqrt{x} + \frac{1}{x^2} \right) dx = \int \left(x^{\frac{1}{2}} + x^{-2} \right) dx = \frac{x^{\frac{3}{2}}}{\frac{3}{2}} + \frac{x^{-1}}{-1} + c = \frac{2}{3}x^{\frac{3}{2}} - x^{-1} + c$$

> There are lots of conversions to indices and back again in this topic. If you are unsure about indices, look back at Topic 1.

Example

① Find $\int \left(x^3+3x^2-2x+1\right) dx$

Answer

① $\int \left(x^3+3x^2-2x+1\right) dx$

$$= \frac{x^4}{4} + \frac{3x^3}{3} - \frac{2x^2}{2} + x + c$$

$$= \frac{x^4}{4} + x^3 - x^2 + x + c$$

Grade boost

Students frequently lose marks because they differentiate instead of integrating. Another way they lose marks is forgetting to include the constant of integration.

Example

② Find $\int \left(3x^2 + \frac{1}{x^2} + \sqrt{x}\right) dx$

Note the reciprocal and the root in this integral. Both must be changed to index form so they can be integrated.

Answer

② $\int \left(3x^2 + \frac{1}{x^2} + \sqrt{x}\right) dx$

$$= \int \left(3x^2 + x^{-2} + x^{\frac{1}{2}}\right) dx$$

Be careful with the signs when dealing with negative indices.

Note that when you divide by a fraction, you invert the fraction and then multiply by it. Hence

$$\frac{x^{\frac{3}{2}}}{\left(\frac{3}{2}\right)} = \frac{2}{3} x^{\frac{3}{2}}$$

$$= \frac{3x^3}{3} + \frac{x^{-1}}{(-1)} + \frac{x^{\frac{3}{2}}}{\left(\frac{3}{2}\right)} + c$$

$$= x^3 - x^{-1} + \frac{2}{3} x^{\frac{3}{2}} + c$$

Example

③ Find $\int \left(4x^3 - \frac{2}{\sqrt{x}}\right) dx$

Answer

③ $\int \left(4x^3 - \frac{2}{\sqrt{x}}\right) dx$

Note that $\dfrac{1}{\sqrt{x}} = x^{-\frac{1}{2}}$

$\int \left(4x^3 - 2x^{-\frac{1}{2}}\right) dx$

$$= \frac{4x^4}{4} - \frac{2x^{\frac{1}{2}}}{\frac{1}{2}} + c$$

Remember to include c, the constant of integration.

$$= x^4 - 4x^{\frac{1}{2}} + c$$

Example

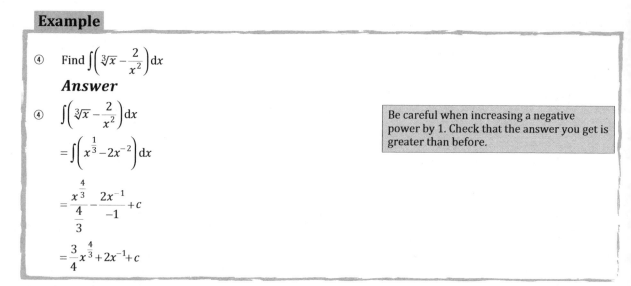

④ Find $\int\left(\sqrt[3]{x}-\dfrac{2}{x^2}\right)dx$

Answer

④ $\int\left(\sqrt[3]{x}-\dfrac{2}{x^2}\right)dx$

$=\int\left(x^{\frac{1}{3}}-2x^{-2}\right)dx$

$=\dfrac{x^{\frac{4}{3}}}{\frac{4}{3}}-\dfrac{2x^{-1}}{-1}+c$

$=\dfrac{3}{4}x^{\frac{4}{3}}+2x^{-1}+c$

> Be careful when increasing a negative power by 1. Check that the answer you get is greater than before.

Approximation of the area under a curve using the trapezium rule

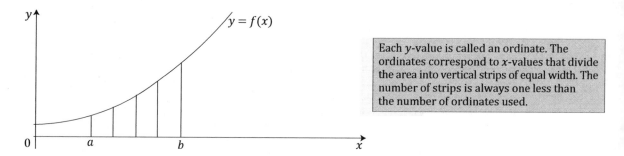

> Each y-value is called an ordinate. The ordinates correspond to x-values that divide the area into vertical strips of equal width. The number of strips is always one less than the number of ordinates used.

Look at the curve above. Suppose we want to find the area under the curve between $x = a$ and $x = b$. The area under the curve between these two points, is divided into strips (4 in this case) of equal width. Each strip can be approximated to a trapezium by making the assumption that there is a straight line at the top of the strip rather than a curve. By working out the areas of all the trapezia (the plural for trapezium) and then adding them together we get the approximate area under the curve. The greater number of strips used, the more accurate the approximation to the true area becomes.

The approximate area under a curve may be found using a formula called the trapezium rule. The trapezium rule is:

$$\int_a^b y\,dx \approx \frac{1}{2}h\left\{(y_0+y_n)+2(y_1+y_2+...+y_{n-1})\right\}, \text{ where } h=\frac{b-a}{n}$$

h is the width of the strips used to estimate the area.

n is the number of strips used. n is always one less than the number of ordinates used. For example if 5 ordinates are being used to estimate the area, then the number of strips used n will be 4.

y_0 and y_n are the first and last ordinates respectively.

$y_1, y_2, ..., y_{n-1}$ are the other ordinates between the first and the last.

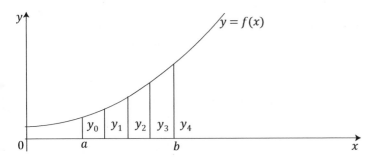

Overestimating and underestimating areas using the trapezium rule

The trapezium rule will overestimate the area if the tops of the trapezia are above the curve and underestimate the area if they are below the curve.

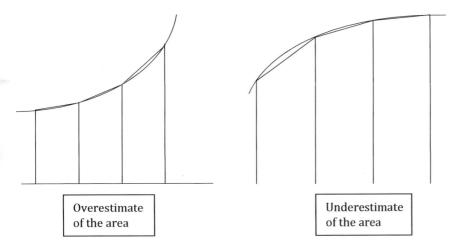

| Overestimate of the area | Underestimate of the area |

Increasing the accuracy of the estimate of the area using the trapezium rule

The accuracy of the estimate of the area obtained using the trapezium rule can be increased by increasing the number of ordinates or strips used.

The use of the trapezium rule is explained using the following examples.

Example

① Use the trapezium rule with five ordinates to find an approximate value for the integral

$$\int_0^1 \sqrt{\frac{1}{1+x^2}}\, dx$$

Show your working and give your answer correct to three decimal places.

Answer

① $h = \dfrac{b-a}{n} = \dfrac{1-0}{4} = 0.25$

> h gives the width of the strips. b and a are the top and bottom limits of the integral. n the number of strips used is one less than the number of ordinates.

$\displaystyle\int_0^1 \sqrt{\dfrac{1}{1+x^2}}\,dx \approx \dfrac{1}{2}h\{(y_0+y_4)+2(y_1+y_2+y_3)\}$

> This formula and also the formula for h are obtained from the formula booklet.

When $x = 0, y_0 = \sqrt{\dfrac{1}{1+(0)^2}} = 1$

> Starting from the lower limit (i.e. the value of a) and working in steps of h (i.e. 0.25 here) the values of x are substitute into the expression inside the integral. This gives the ordinates y_0, y_1 etc. which can then be entered into the formula for the Trapezium Rule.

$x = 0.25, y_1 = \sqrt{1+\dfrac{1}{(0.25)^2}} = 4.12311$

$x = 0.5, y_2 = \sqrt{1+\dfrac{1}{(0.5)^2}} = 2.23607$

$x = 0.75, y_3 = \sqrt{1+\dfrac{1}{(0.75)^2}} = 2.77778$

> Notice the question asks to give the answer correct to three decimal places. You must give the intermediate working to more decimal places and then give the final answer to three decimal places.

$x = 1, y_4 = \sqrt{1+\dfrac{1}{(1)^2}} = 1.41421$

$\displaystyle\int_0^1 \sqrt{\dfrac{1}{1+x^2}}\,dx \approx \dfrac{1}{2}\times 0.25\,\{(1+1.41421)+2(4.12311+2.23607+2.77778)\}$

≈ 2.58602

≈ 2.586 (to 3 decimal places)

Example

② Use the trapezium rule with five ordinates to find an approximate value for the integral $\displaystyle\int_1^2 \sqrt{4+x^3}\,dx$

Show your working and give your answer correct to three decimal places. [4]

(WJEC C2 Jan 2011 Q1)

Answer

② $h = \dfrac{b-a}{n} = \dfrac{2-1}{4} = 0.25$

> It is important to note that n is the number of strips and not the number of ordinates. Here there are 5 ordinates so there will be 4 strips hence $n = 4$.

This means that you start at the value of a (i.e. 1 in this case) and go up in steps of h (0.25 here) until the value of b is reached (2 in this case). These are the values of x which are best tabulated in the following way:

When $x = 1, y_0 = \sqrt{4+1^3} = \sqrt{5} = 2.23607$

$x = 1.25, y_1 = \sqrt{4+1.25^3} = 2.43990$

$x = 1.50, y_2 = \sqrt{4+1.50^3} = 2.71570$

$x = 1.75, y_3 = \sqrt{4+1.75^3} = 1.66667$

$x = 2, y_4 = \sqrt{4+2^3} = 3.46410$

> Always work to at least one decimal place beyond that required for the answer.

$\displaystyle\int_a^b y\,dx \approx \dfrac{1}{2}h\{(y_0+y_n)+2(y_1+y_2+...+y_{n-1})\}$

$\displaystyle\int_1^2 \sqrt{4+x^3}\,dx$

> The formula for the trapezium rule is obtained from the formula booklet. You do not have to remember it.

$$\approx \frac{1}{2} \times 0.25 \{(2.23607 + 3.46410) + 2(2.43990 + 2.71570 + 1.66667)\}$$

≈ 2.30824

≈ 2.308 (to 3 decimal places)

> **≫ Grade boost**
>
> Remember to give your answer to the required number of decimal places or significant figures specified in the question.

Interpretation of the definite integral as the area under a curve

Integrals in the form $\int_a^b y \, dx$ are called definite integrals because the result will be a definite answer, usually a number, with no constant of integration.

The definite integral is found by substituting the limits into the result of the integration, and subtracting the value corresponding to the lower limit from the value corresponding to the upper limit.

Definite integrals represent the area under the curve $y = f(x)$ between the two x-values $x = a$ and $x = b$.

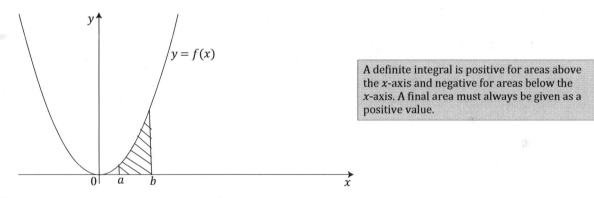

> A definite integral is positive for areas above the x-axis and negative for areas below the x-axis. A final area must always be given as a positive value.

Evaluation of definite integrals

Definite integrals have a numerical value which is obtained when the two limits are substituted for x into the result of the integration as the following examples show.

Example

① Find $\int_1^2 (3x^2 - x + 4) \, dx$

Answer

① $\int_1^2 (3x^2 - x + 4) \, dx$

$$= \left[\frac{3x^3}{3} - \frac{x^2}{2} + 4x \right]_1^2$$

> Once you have integrated, put square brackets around the result and write the limits as shown here.

$$= \left[x^3 - \frac{x^2}{2} + 4x \right]_1^2$$

$$= \left[\left(2^3 - \frac{2^2}{2} + 4(2) \right) - \left(1^3 - \frac{1^2}{2} + 4(1) \right) \right]$$

$$= \left[(8-2+8) - \left(1 - \frac{1}{2} + 4 \right) \right]$$

$$= 14 - 4\frac{1}{2} = 9\frac{1}{2} \text{ square units}$$

Two pairs of brackets are used. The first contains the top limit substituted in for x. The second contains the bottom limit substituted in for x. The contents of the second bracket are subtracted from the contents of the first.

Example

② (a) Find $\int \left(\frac{2}{\sqrt{x}} - x^3 + \frac{2}{x^2} \right) dx$

(b)

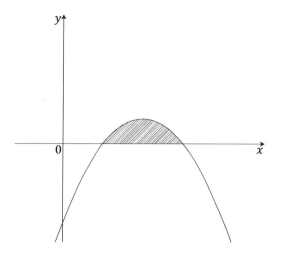

The diagram shows a sketch of the curve $y = (1 - x)(x - 3)$.

(i) Find the coordinates of the points of intersection of the curve and the x-axis.

(ii) Find the area of the shaded region.

$x - 3 - x^2 + 3x$

$-x^2 + 4x - 3$

Change to indices so that the expression can be integrated.

Answer

② (a) $\int \left(\frac{2}{\sqrt{x}} - x^3 + \frac{2}{x^2} \right) dx$

$\int \left(2x^{-\frac{1}{2}} - x^3 + 2x^{-2} \right) dx$

$$= \frac{2x^{\frac{1}{2}}}{\frac{1}{2}} - \frac{x^4}{4} + \frac{2x^{-1}}{-1} + c$$

$$= 4x^{\frac{1}{2}} - \frac{x^4}{4} - 2x^{-1} + c$$

Grade boost

Forgetting to include the constant of integration frequently costs students marks.

(b) (i) When $y = 0$, $(1-x)(x-3) = 0$

Giving $x = 1$ or $x = 3$

Put the equation of the curve equal to 0 to find the points of intersection with the x-axis.

Coordinates of the points of intersection with the x-axis are $(1, 0)$ and $(3, 0)$.

(ii) Shaded area $= \int_1^3 y \, dx = \int_1^3 (1-x)(x-3) \, dx$

$$= \int_1^3 (-3 + 4x - x^2) \, dx$$

$$= \left[-3x + 2x^2 - \frac{x^3}{3} \right]_1^3$$

$$= \left[(-9 + 18 - 9) - \left(-3 + 2 - \frac{1}{3} \right) \right]$$

$$= 0 - \left(-\frac{4}{3} \right)$$

$$= \frac{4}{3} \text{ square units}$$

Example

③ (a) Find $\int \left(5\sqrt{x} - \frac{4}{x^{\frac{2}{3}}} \right) dx$ [2]

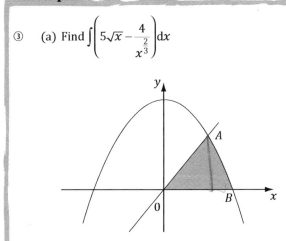

(b) The diagram shows a sketch of the curve $y = 4 - x^2$ and the line $y = 3x$. The curve and the line intersect at the point A in the first quadrant and the curve intersects the positive x-axis at the point B.

(i) Showing your working, find the coordinates of A and the coordinates of B.

(ii) Find the area of the shaded region. [12]

(WJEC C2 May 2008 Q6)

Answer

③ (a) $\int \left(5\sqrt{x} - \dfrac{4}{x^{\frac{2}{3}}} \right) dx$

$= \int \left(5x^{\frac{1}{2}} - 4x^{-\frac{2}{3}} \right) dx$

$= \dfrac{5x^{\frac{3}{2}}}{\frac{3}{2}} - \dfrac{4x^{\frac{1}{3}}}{\frac{1}{3}} + c$

$= \dfrac{2}{3} \times 5x^{\frac{3}{2}} - 3 \times 4x^{\frac{1}{3}} + c$

$= \dfrac{10}{3}x^{\frac{3}{2}} - 12x^{\frac{1}{3}} + c$

> Remember that to integrate you increase the index by one and then divide by the new index.

> Remember when you divide by a fraction, turn the fraction upside down and multiply by the new fraction, so dividing by $\dfrac{1}{3}$ is the same as multiplying by 3.

> As this is an indefinite integral you must remember to include the constant c.

(b) Equating the y-values gives

$3x = 4 - x^2$

$x^2 + 3x - 4 = 0$

Factorising gives $(x - 1)(x + 4) = 0$

Solving gives $x = 1$ or -4

The x-coordinate of A cannot be -4 as A is in the first quadrant.

As $y = 3x$, substituting $x = 1$ into this gives $y = 3(1) = 3$

Hence A is the point $(1, 3)$

For the coordinates of B, substitute $y = 0$ into the equation $y = 4 - x^2$

So, $0 = 4 - x^2$

$x^2 = 4$

$x = \pm 2$

From the question, B has a positive x-value, so $x = 2$.

Hence B has coordinates $(2, 0)$

> To find the coordinates of A, solve the equation of the curve simultaneously with that of the straight line.

> Always look at the diagram to check the significance of the points found.

> Remember to include both values of x here.

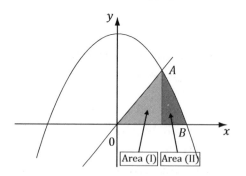

Area of (I) = area of a triangle $= \dfrac{1}{2} \times base \times height$

$= \dfrac{1}{2} \times 1 \times 3 = \dfrac{3}{2}$

Area (II) under the curve $= \int_1^2 \left(4 - x^2\right) dx$

$= \left[\left(4x - \dfrac{x^3}{3} \right) \right]_1^2$

$= \left(8 - \dfrac{8}{3} \right) - \left(4 - \dfrac{1}{3} \right)$

$= 4 - 2\dfrac{1}{3}$

$= 1\dfrac{2}{3}$

Total area = area (I) + area (II) $= \dfrac{3}{2} + 1\dfrac{2}{3} = 3\dfrac{1}{6}$ square units

Examination style questions

①

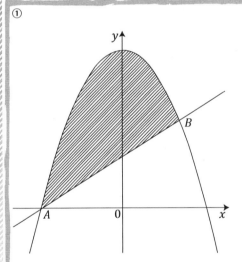

The diagram shows a sketch of the curve $y = 9 - x^2$ and the line $y = x + 3$.

The line and the curve intersect at the points A and B.

(a) Find the coordinates of A and B. [4]

(b) Find the area of the shaded region. [7]

Answer

① (a) Solving the equations of the curve and straight line simultaneously to find the coordinates of the points of intersection A and B.

Equating the y-values gives:

$9 - x^2 = x + 3$

$x^2 + x - 6 = 0$

> Form a quadratic equation, then factorise and finally solve it.

$(x + 3)(x - 2) = 0$

Solving gives $x = -3$ or 2

Substitute both values of x into the equation of the straight line to find the corresponding y-coordinates.

When $x = -3$, $y = (-3) + 3 = 0$

> Looking at the diagram this is point A.

When $x = 2$, $y = 2 + 3 = 5$

By looking at the graph A is $(-3, 0)$ and B is $(2, 5)$.

(b)   Area under the curve between $x = -3$ and $x = 2$ is given by

$$\int_{-3}^{2}\left(9-x^2\right)dx$$

$$=\left[\left(9x-\frac{x^3}{3}\right)\right]_{-3}^{2}$$

$$=\left[\left(9(2)-\frac{(2)^3}{3}\right)-\left(9(-3)-\frac{(-3)^3}{3}\right)\right]$$

$$=\left[\left(18-\frac{8}{3}\right)-(-27+9)\right]$$

$$=33\frac{1}{3}$$

Area of the triangle under line $y = x + 3$ is $\frac{1}{2}\times5\times5 = 12\frac{1}{2}$

Hence, shaded area $= 33\frac{1}{3}-12\frac{1}{2} = 20\frac{5}{6}$ square units

Test yourself

Answer the following questions and check your answers.

①   Find $\int\left(4x^{\frac{1}{3}} - \frac{2}{\sqrt[3]{x}}\right)dx$

②   Find $\int\left(\sqrt[3]{x} - \frac{1}{x^4}\right)dx$

③   Find $\int\left(\frac{4}{x^3} - 6x^{\frac{1}{5}}\right)dx$

④   Find $\int\left(\frac{2}{\sqrt{x}} - x^{\frac{3}{2}}\right)dx$

⑤   Find $\int_{0}^{4}\left(x^{-\frac{1}{2}} + 2x\right)dx$

⑥   Use the trapezium rule with five ordinates to find an approximate value for the integral

$$\int_{0}^{4}\left(\frac{1}{1+\sqrt{x}}\right)dx$$

Show your workings and give your answer correct to three decimal places.

(Note: answers to Test yourself are found at the back of the book.)

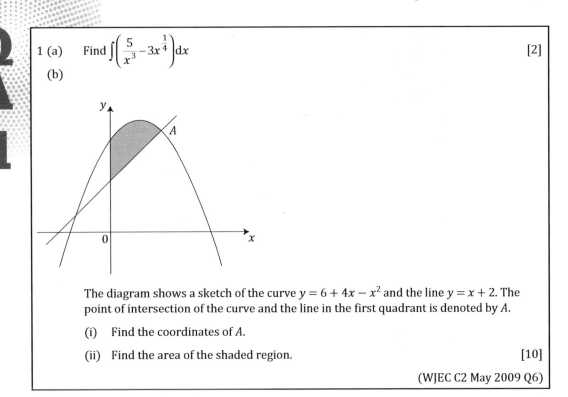

1 (a)   Find $\int\left(\dfrac{5}{x^3}-3x^{\frac{1}{4}}\right)dx$ [2]

(b)

The diagram shows a sketch of the curve $y = 6 + 4x - x^2$ and the line $y = x + 2$. The point of intersection of the curve and the line in the first quadrant is denoted by A.

(i)   Find the coordinates of A.

(ii)   Find the area of the shaded region.   [10]

(WJEC C2 May 2009 Q6)

Answer

1 (a) $\int\left(\dfrac{5}{x^3}-3x^{\frac{1}{4}}\right)dx$

> Remember to change the reciprocal to a negative power on the top before integrating.

$= \int\left(5x^{-3}-3x^{\frac{1}{4}}\right)dx$

> As this is an indefinite integral the constant of integration, c must be included in the answer.

$= \dfrac{5x^{-2}}{-2} - \dfrac{3x^{\frac{5}{4}}}{\frac{5}{4}} + c$

$= -\dfrac{5}{2}x^{-2} - \dfrac{12}{5}x^{\frac{5}{4}} + c$

(b) (i) Solving the equations of the curve and the straight line simultaneously by equating the y-values gives

$6 + 4x - x^2 = x + 2$

$0 = x^2 - 3x - 4$

$0 = (x + 1)(x - 4)$

> Equating the y-values will produce an equation which can be solved to find the x-coordinates of the two points of intersection of the curve with the straight line.

Giving $x = -1$ or $x = 4$

From the question, point A is in the first quadrant.

> You need to decide which of these x-coordinates is that of point A.

Hence $x = 4$

When $x = 4, y = 4 + 2 = 6$

> Don't forget that you need both coordinates for point A.

So A is the point $(4, 6)$

(ii) Area under the curve between $x = 0$ and $x = 4$ is given by

$$\int_0^4 y \, dx = \int_0^4 (6 + 4x - x^2) \, dx = \left[6x + 2x^2 - \frac{x^3}{3} \right]_0^4$$

$$= \left[\left(24 + 32 - \frac{64}{3} \right) - (0) \right]$$

$$= 34\frac{2}{3} \text{ square units}$$

When $x = 4$, $y = 6$

To find the y-value of the point where the straight line cuts the y-axis, substituting $x = 0$ into the equation of the straight line:

$y = x + 2 = 0 + 2 = 2$

Area of trapezium formed under the straight line between $x = 0$ and $x = 4$

$= \dfrac{1}{2}$ (sum of the two parallel sides) $\times$ (distance between the parallel sides)

$= \dfrac{1}{2} \times (2 + 6) \times 4 = 16$ square units

Shaded area $= 34\dfrac{2}{3} - 16 = 18\dfrac{2}{3}$ square units

> Remember that an answer on its own with no working will earn no marks.

> An alternative method to find the area of the trapezium (i.e. the area under the line $y = x + 2$) would be to integrate y between the limits 0 and 4, i.e.
> $$\int_0^4 x + 2 \, dx = \left[\frac{x^2}{2} + 2x \right]_0^4 = \frac{16}{2} + 8 = 16$$

Q&A 2

2 (a)   Find $\displaystyle\int \left(\frac{3}{x^2} - 2\sqrt{x} \right) dx$ [2]

(b)

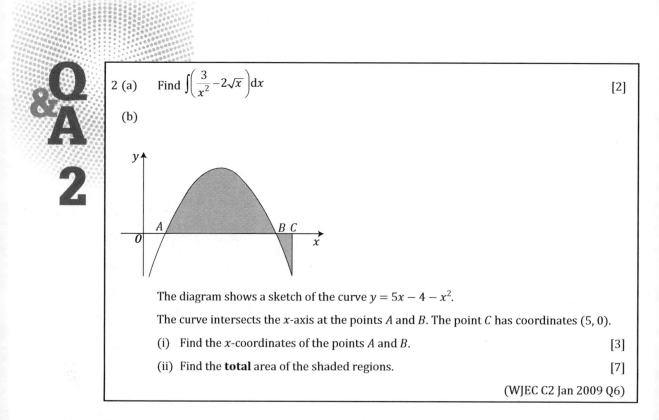

The diagram shows a sketch of the curve $y = 5x - 4 - x^2$.

The curve intersects the x-axis at the points A and B. The point C has coordinates $(5, 0)$.

(i)   Find the x-coordinates of the points A and B. [3]

(ii) Find the **total** area of the shaded regions. [7]

(WJEC C2 Jan 2009 Q6)

Answer

2 (a) $\int\left(\dfrac{3}{x^2}-2\sqrt{x}\right)dx$

$= \int\left(3x^{-2}-2x^{\frac{1}{2}}\right)dx$

$= \dfrac{3x^{-1}}{-1}-\dfrac{2x^{\frac{3}{2}}}{\frac{3}{2}}+c$

$= -3x^{-1}-\dfrac{4}{3}x^{\frac{3}{2}}+c$

> **Grade boost**
>
> You must remember to include a constant of integration. A mark is withheld if the constant of integration is not included in the answer for an indefinite integral.

(b) (i) For the coordinates of A and B substituting $y = 0$ gives

$0 = 5x - 4 - x^2$

$x^2 - 5x + 4 = 0$

$(x - 4)(x - 1) = 0$

$x = 1$ or $x = 4$

At point A, $x = 1$ and at point B, $x = 4$

(ii) Area under curve between A and $B = \int_{1}^{4}\left(5x - 4 - x^2\right)dx$

$= \left[\dfrac{5x^2}{2} - 4x - \dfrac{x^3}{3}\right]_{1}^{4}$

$= \left[\left(\dfrac{5\times16}{2} - 16 - \dfrac{64}{3}\right) - \left(\dfrac{5\times1}{2} - 4 - \dfrac{1}{3}\right)\right]$

$= \left[\left(2\dfrac{2}{3}\right) + \left(1\dfrac{5}{6}\right)\right]$

$= 4\dfrac{1}{2}$ square units

Area above curve between B and C

$= -\int_{4}^{5}\left(5x - 4 - x^2\right)dx$

$= -\left[\dfrac{5x^2}{2} - 4x - \dfrac{x^3}{3}\right]_{4}^{5}$

$= -\left[\left(\dfrac{5\times25}{2} - 20 - \dfrac{125}{3}\right) - \left(\dfrac{5\times16}{2} - 16 - \dfrac{64}{3}\right)\right]$

$= -62\dfrac{1}{2} + 20 + \dfrac{125}{3} + 40 - 16 - \dfrac{64}{3}$

$= \dfrac{11}{6}$ square units

> Using integration to find the area between a curve and the x-axis will give a positive value for regions above the x-axis and a negative value for regions below the x-axis. In order find the total shaded area, positive areas need to be added together. To ensure that areas are positive, the value of definite integrals will need to be negated for regions below the axis.

Total area $= 4\dfrac{1}{2} + \dfrac{11}{6} = \dfrac{19}{3} = 6\dfrac{1}{3}$ square units

3   Use the trapezium rule with five ordinates to find an approximate value for the integral

$$\int_1^2 \sqrt{1+\frac{1}{x}}\, dx$$

Show your working and give your answer correct to three decimal places.

(WJEC C2 May 2010 Q1)

Answer

3    $h = \dfrac{b-a}{n} = \dfrac{2-1}{4} = 0.25$

> An answer on its own with no working will earn no marks.

$$\int_1^2 \sqrt{1+\frac{1}{x}}\, dx \approx \frac{1}{2} h \left\{ (y_0 + y_n) + 2(y_1 + y_2 + \ldots + y_{n-1}) \right\}$$

When $x = 1$, $y_0 = \sqrt{1+\dfrac{1}{1}} = \sqrt{2}, = 1.41421$

$x = 1.25$, $y_1 = \sqrt{1+\dfrac{1}{1.25}} = 1.34164$

$x = 1.5$, $y_2 = \sqrt{1+\dfrac{1}{1.5}} = 1.29099$

$x = 1.75$, $y_3 = \sqrt{1+\dfrac{1}{1.75}} = 1.25357$

$x = 2$, $y_n = \sqrt{1+\dfrac{1}{2}} = 1.22474$

Substituting these values into the formula gives

$$\int_1^2 \sqrt{1+\frac{1}{x}}\, dx \approx \frac{1}{2} \times 0.25 \left\{ (1.41421 + 1.22474) + 2(1.34164 + 1.29099 + 1.25357) \right\}$$

≈ 1.30142

≈ 1.301 (to 3 decimal places)

Summary C2 Pure Mathematics

1 Sequences, arithmetic series and geometric series

The nth term of an arithmetic sequence

nth term $t_n = a + (n-1)d$

where a is the first term, d is the common difference and n is the number of terms.

The sum to n terms of an arithmetic series

$$S_n = \frac{n}{2}(2a + (n-1)d)$$

The nth term of a geometric sequence

nth term $t_n = ar^{n-1}$

where a is the first term, r is the common ratio and n is the number of terms.

The sum to n terms of a geometric series

$$S_n = \frac{a(1-r^n)}{1-r} \text{ provided } r \neq 1$$

The sum to infinity of a geometric series

$$S_\infty = \frac{a}{1-r}$$

Note that for the sum to infinity to exist $|r| < 1$

2 Logarithms and their uses

The logarithm and exponential functions

A logarithm of a positive number to base a is the power to which the base must be raised in order to give the positive number.

$y = a^x$

$\log_a y = x$

These two equations have the same meaning and you should be able to convert readily between them.

Some important results

For a positive base a, the following are true:

$\log_a a = 1$, as $a^1 = a$

$\log_a 1 = 0$, as $a^0 = 1$

The three laws of logarithms

$\log_a x + \log_a y = \log_a (x\,y)$

$\log_a x - \log_a y = \log_a \dfrac{x}{y}$

$\log_a x^k = k \log_a x$

3 Coordinate geometry of the circle

The two forms for the equation of a circle

A circle having an equation in the form

$(x - a)^2 + (y - b)^2 = r^2$ has centre (a, b) and radius r.

A circle having an equation in the form

$x^2 + y^2 + 2gx + 2fy + c = 0$ has centre $(-g, -f)$ and radius $\sqrt{g^2 + f^2 - c}$

Circle properties

The angle in a semicircle is a right angle.

The perpendicular from the centre of a circle to a chord bisects the chord.

The radius to a point on the circle and the tangent through the same point are perpendicular to each other.

Summary of the Core 1 material needed for this topic

The gradient of the line joining points (x_1, y_1) and (x_2, y_2) is given by:

$$\text{Gradient} = \frac{y_2 - y_1}{x_2 - x_1}$$

For two lines to be parallel to each other, they must have the same gradient.

When two lines are perpendicular to each other (i.e. they make an angle of 90°), the product of their gradients is −1.

If one line has a gradient m_1 and the other a gradient of m_2 then $m_1 m_2 = -1$.

The equation of a straight line with gradient m and which passes through a point (x_1, y_1) is given by:

$$y - y_1 = m(x - x_1)$$

The length of a straight line joining the two points (x_1, y_1) and (x_2, y_2) is given by:

$$\sqrt{(x_2 - x_1)^2 + (y_2 - y_1)^2}$$

The mid-point of a line joining the points (x_1, y_1) and (x_2, y_2) is given by:

$$\left(\frac{x_1 + x_2}{2}, \frac{y_1 + y_2}{2} \right)$$

4 Trigonometry

The sine and cosine rules and the formula for the area of a triangle

The sine rule states: $\dfrac{a}{\sin A} = \dfrac{b}{\sin B} = \dfrac{c}{\sin C}$

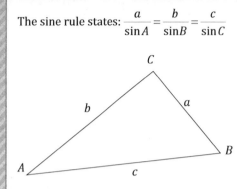

The cosine rule states: $a^2 = b^2 + c^2 - 2bc \cos A$

Area of triangle $= \dfrac{1}{2} ab \sin C$

Radian measure, arc length, area of sector and area of segment

π radians $= 180°$ 2π radians $= 360°$

$\dfrac{\pi}{2}$ radians $= 90°$ $\dfrac{\pi}{4}$ radians $= 45°$

$\dfrac{\pi}{3}$ radians $= 60°$ $\dfrac{\pi}{6}$ radians $= 30°$

The length of an arc making an angle of θ radians at the centre $l = r\theta$

Area of sector making an angle of θ radians at the centre $= \dfrac{1}{2} r^2 \theta$

Area of segment $= \dfrac{1}{2} r^2 (\theta - \sin \theta)$

Trigonometric relationships

$\tan \theta = \dfrac{\sin \theta}{\cos \theta}$

$\cos^2 \theta + \sin^2 \theta = 1$

5 Integration

Indefinite integration is the reverse process of differentiation. When integrating indefinitely you must remember to include the constant of integration.

$$\int x^n \, dx = \frac{x^{n+1}}{n+1} + c \text{ (provided } n \neq -1)$$

A definite integral is positive for areas above the x-axis and negative for areas below the x-axis.

A final area must always be given as a positive value.

The trapezium rule can be used for estimating areas.

$$\int_a^b y \, dx \approx \frac{1}{2} h \{(y_0 + y_n) + 2(y_1 + y_2 + \ldots y_{n-1})\} \text{, where } h = \frac{b-a}{n}$$

Test yourself answers

Core 1 Pure Mathematics 1

1 Indices and surds

① $\quad y = 5x^{\frac{1}{2}} + 45x^{-1} - 7$

② (a) $\quad 1$

   (b) $\quad \dfrac{1}{9}$

   (c) $\quad 2$

   (d) $\quad \dfrac{1}{5}$

   (e) $\quad 64$

③ (a) $\quad \sqrt{48} + \dfrac{12}{\sqrt{3}} - \sqrt{27} = 4\sqrt{3} + \dfrac{12\sqrt{3}}{\sqrt{3}\sqrt{3}} - 3\sqrt{3} = 4\sqrt{3} + 4\sqrt{3} - 3\sqrt{3} = 5\sqrt{3}$

   (b) $\quad \dfrac{2+\sqrt{5}}{3+\sqrt{5}} = \dfrac{\left(2+\sqrt{5}\right)\left(3-\sqrt{5}\right)}{\left(3+\sqrt{5}\right)\left(3-\sqrt{5}\right)} = \dfrac{6+\sqrt{5}-5}{9-5} = \dfrac{1+\sqrt{5}}{4}$

2 Quadratic functions, equations, graphs and their transformations

① As this question is about the nature of roots, we first find the discriminant

$b^2 - 4ac = (5)^2 - 4(k)\,(-7) = 25 + 28k$

For no real roots, $b^2 - 4ac < 0$

Hence $25 + 28k < 0$

So, $28\,k < -25$, giving $k < -\dfrac{25}{28}$

② $x^2 - 6x + 8 > 0$

$(x-4)(x-2) > 0$

As the curve $y = x^2 - 6x + 8$ has a positive coefficient of x^2 the curve will be $\cup$ shaped cutting the x-axis at $x = 4$ and $x = 2$.

Sketching the curve for $y = x^2 - 6x + 8$ gives the following:

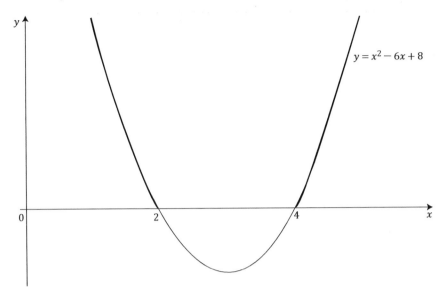

We want the part of the graph which is above the x-axis.

The range of values for which this occurs is $x < 2$ and $x > 4$.

③ $5x^2 - 20x + 10$

$= 5[x^2 - 4x + 2]$

$= 5[(x - 2)^2 - 4 + 2]$

$= 5(x - 2)^2 - 10$

Giving $a = 5$, $b = -2$ and $c = -10$

> Remember that the coefficient of x^2 needs to be 1 before you complete the square. Here it is necessary to take a 5 out of the square bracket as a factor.

④ $1 - 3x < x + 7$

$-3x < x + 6$

$-4x < 6$

$x > -\dfrac{6}{4}$

$x > -\dfrac{3}{2}$

> The inequality sign is reversed because both sides have been divided by a negative quantity (i.e. –4).

≫ Grade boost

If you do not cancel fractions you may lose marks.

⑤ $y = x + 4$ and $y = x^2 - 7x + 20$

$x + 4 = x^2 - 7x + 20$

$x^2 - 8x + 16 = 0$

$(x - 4)(x - 4) = 0$

$(x - 4)^2 = 0$

> If the line and the curve touch then the resulting equation will have a repeated root.

There is one repeated solution to this equation which proves that the straight line and curve touch.

Solving gives $x = 4$

Substituting $x = 4$ into the equation of the straight line

There is only one solution to the quadratic which means the straight line and curve touch at only one point.

$y = 4 + 4 = 8$

Hence, the coordinates of the point of contact are $(4, 8)$

An alternative method for proving that the curve and straight line touch at one point is to find the discriminant and show that it equals zero.

For example the equation $x^2 - 8x + 16 = 0$ has discriminant $b^2 - 4ac = (-8)^2 - 4(1)(16) = 64 - 64 = 0$. This shows there are two real equal roots showing the curve and straight line touch at a single point.

3 Coordinate geometry

① (a)   Gradient of $AB = \dfrac{y_2 - y_1}{x_2 - x_1} = \dfrac{1 - 0}{4 - 1} = \dfrac{1}{3}$

Gradient of $CD = \dfrac{y_2 - y_1}{x_2 - x_1} = \dfrac{4 - 3}{2 - (-1)} = \dfrac{1}{3}$

As the gradients of AB and CD are the same the two lines are parallel.

(b)   Gradient of $AB = \dfrac{1}{3}$ and AB passes through point A $(1, 0)$ so equation of AB is:

$y - y_1 = m(x - x_1)$

$y - 0 = \dfrac{1}{3}(x - 1)$

$3y = x - 1$

Rearranging this equation so that it is in the form asked for by the question gives:

$x - 3y - 1 = 0$

② (a)   Gradient of $AB = \dfrac{y_2 - y_1}{x_2 - x_1} = \dfrac{-1 - 4}{k - (-7)} = \dfrac{-5}{k + 7}$

But gradient of $AB = -\dfrac{1}{2}$ so

$\dfrac{-5}{k + 7} = -\dfrac{1}{2}$

The equation is multiplied through by the common denominator, $2(k + 7)$.

$-5 \times 2 = -1(k + 7)$

$-10 = -k - 7$

Giving $k = 3$.

(b)   The product of the gradients of perpendicular lines is -1. Hence,

$m\left(-\dfrac{1}{2}\right) = -1$

Hence gradient of $BC = 2$

Equation of BC is:

$y - y_1 = m(x - x_1)$ where $m = 2$ and $(x_1, y_1) = (3, -1)$.

$y - (-1) = 2(x - 3)$

$y + 1 = 2x - 6$

$2x - y - 7 = 0$

③ (a) Gradient of $AB = \dfrac{y_2 - y_1}{x_2 - x_1} = \dfrac{6 - 2}{1 - (-3)} = \dfrac{4}{4} = 1$

Gradient of $BC = \dfrac{y_2 - y_1}{x_2 - x_1} = \dfrac{1 - 6}{6 - 1} = \dfrac{-5}{5} = -1$

Product of gradients $= (1)(-1) = -1$ proving that the two lines are perpendicular to each other.

(b) $\sqrt{(x_2 - x_1)^2 + (y_2 - y_1)^2}$

Substituting the coordinates $A\ (-3, 2)$ and $B\ (1, 6)$ into the formula gives

$AB = \sqrt{(1 - (-3))^2 + (6 - 2)^2} = \sqrt{16 + 16} = \sqrt{32}$ units

Using the coordinates $B\ (1, 6)$ and $C\ (6, 1)$ in the formula gives

$BC = \sqrt{(6 - 1)^2 + (1 - 6)^2} = \sqrt{25 + 25} = \sqrt{50}$ units

(c) $\operatorname{Tan} A\hat{C}B = \dfrac{AB}{BC} = \dfrac{\sqrt{32}}{\sqrt{50}} = \dfrac{\sqrt{16 \times 2}}{\sqrt{25 \times 2}} = \dfrac{4\sqrt{2}}{5\sqrt{2}} = \dfrac{4}{5}$

> **≫ Grade boost**
>
> This formula needs to be remembered. If you forget it you can plot the two points on a sketch graph, form a triangle and use Pythagoras theorem to find the length of the hypotenuse.

> Core 1 material is needed here to enable the surds to be simplified.

4 Polynomials and the binomial expansion

① Let $f(x) = 4x^3 + 3x^2 - 3x + 1$

$f(-1) = 4(-1)^3 + 3(-1)^2 - 3(-1) + 1 = 3$

Therefore remainder $= 3$

② (a) Let $f(x) = x^3 + 6x^2 + ax + 6$.

$f(-2) = (-2)^3 + 6(-2)^2 + a(-2) + 6 = 22 - 2a$

If $x + 2$ is a factor, $f(-2) = 0$

Hence, $22 - 2a = 0$

So $a = 11$

(b) $(x + 2)(ax^2 + bx + c) = x^3 + 6x^2 - 11x + 6$

Equating coefficients of x^3 gives $a = 1$.

Equating coefficients of x^2 gives $b + 2a = 6$ and since $a = 1$ this gives $b = 4$.

Equating constant terms gives $2c = 6$ so $c = 3$.

$x^3 + 6x^2 + 11x + 6 = (x + 2)(x^2 + 4x + 3)$

Now $(x + 2)(x^2 + 4x + 3) = 0$

So $(x + 2)(x + 1)(x + 3) = 0$

Solving gives $x = -2, -1$ or -3.

③ (a)(i)　$f(-2) = (-2)^3 - (-2)^2 - 4(-2) + 4 = -8 - 4 + 8 + 4 = 0$

　　(ii)　As there is no remainder, $(x + 2)$ is a factor of $f(x) = x^3 - x^2 - 4x + 4$.

　(b)　$x^3 - x^2 - 4x + 4 = (x + 2)(ax^2 + bx + c)$

　　　Equating coefficients of x^3 gives $a = 1$.

　　　Equating coefficients of x^2 gives $-1 = b + 2a$ so $-1 = b + 2$.

　　　Hence $b = -3$.

　　　Equating constant terms gives $4 = 2c$ so $c = 2$.

　　　Substituting these values in for a, b and c gives

　　　$x^3 - x^2 - 4x + 4 = (x + 2)(x^2 - 3x + 2)$

　　　$= (x + 2)(x - 2)(x - 1)$

　　　Now $f(x) = 0$ so $(x + 2)(x - 2)(x - 1) = 0$

　　　Solving gives $x = -2, 2$ or 1

④ Obtaining the formulae from the formula booklet

$(a + b)^n = a^n + \binom{n}{1}a^{n-1}b + \binom{n}{2}a^{n-2}b^2 + \ldots$

$\binom{n}{r} = \dfrac{n!}{r!(n-r)!}$

The term in x^2 is given by:

$\binom{n}{2}a^{n-2}b^2$

> The value of $\binom{5}{2}$ could be found using Pascal's triangle by looking along the row which starts at 1 and then 5. The third number in the row (i.e. 10) gives the value of $\binom{5}{2}$.

Here $a = 2$, $b = 3x$ and $n = 5$

So the term in x^2 is $\dfrac{5!}{2!(5-2)!}(2)^3(3x)^2 = 10 \times 8 \times 9x^2 = 720x^2$

Hence, the coefficient of x^2 is 720

⑤ From the formula booklet:

$(1 + x)^n = 1 + nx + \dfrac{n(n-1)}{2!}x^2 + \dfrac{n(n-1)(n-2)}{3!}x^3 + \ldots$

In this case we substitute $3x$ for x and 6 for n.

Hence $(1 + 3x)^6 \approx 1 + (6)(3x) + \dfrac{(6)(5)}{2 \times 1}(3x)^2 + \dfrac{(6)(5)(4)}{3 \times 2 \times 1}(3x)^3$

$\approx 1 + 18x + 135x^2 + 540x^3$

⋙ Grade boost

Here you were asked to simplify the first four terms. Only do what the question asks. Working out more terms will waste your time.

5 Differentiation

① (a)　$y = 4x^2 + 2x - 1$

Increasing x by δx and y by δy gives

$y + \delta y = 4(x + \delta x)^2 + 2(x + \delta x) - 1$

$y + \delta y = 4(x^2 + 2x\delta x + (\delta x)^2) + 2x + 2\delta x - 1$

$y + \delta y = 4x^2 + 8x\delta x + 4(\delta x)^2 + 2x + 2\delta x - 1$

But $y = 4x^2 + 2x - 1$

Subtracting these equations gives

$\delta y = 8x\delta x + 4(\delta x)^2 + 2\delta x$

Dividing both sides by δx

$\dfrac{\delta y}{\delta x} = 8x + 4\delta x + 2$

Letting $\delta x \to 0$

$\dfrac{dy}{dx} = \underset{\delta x \to 0}{\text{limit}} \dfrac{\delta y}{\delta x} = 8x + 2$

(b)　$y = \dfrac{8}{x^2} + 5\sqrt{x} + 1$

Putting this into index form gives $y = 8x^{-2} + 5x^{\frac{1}{2}} + 1$

Differentiating gives $\dfrac{dy}{dx} = -16x^{-3} + \dfrac{5}{2}x^{-\frac{1}{2}}$

This can be written as $\dfrac{dy}{dx} = -\dfrac{16}{x^3} + \dfrac{5}{2\sqrt{x}}$

> It is easier to revert back to reciprocals and roots so that the numbers can be substituted into the derivative. This enables the numerical value of the gradient to be found.

Substituting $x = 1$ gives $\dfrac{dy}{dx} = -\dfrac{16}{1^3} + \dfrac{5}{2\sqrt{1}} = -13.5$

Gradient of curve when $x = 1$ is -13.5

② (a)　$y = 4\sqrt{x} + \dfrac{32}{x} - 3$

$y = 4x^{\frac{1}{2}} + 32x^{-1} - 3$

$\dfrac{dy}{dx} = 2x^{-\frac{1}{2}} - 32x^{-2}$

$\dfrac{dy}{dx} = \dfrac{2}{\sqrt{x}} - \dfrac{32}{x^2}$

When $x = 4$, $\dfrac{dy}{dx} = \dfrac{2}{\sqrt{4}} - \dfrac{32}{4^2} = 1 - 2 = -1$

(b)　Gradient of the tangent at $x = 4$ is -1.

Now $m_1 m_2 = -1$

Gradient of normal $= m$ so $m(-1) = -1$, hence $m = 1$

To find the y-coordinate of the point on the curve where $x = 4$ we substitute $x = 4$ into the equation of the curve.

$$y = 4\sqrt{4} + \frac{32}{4} - 3 = 8 + 8 - 3 = 13$$

Equation of normal having gradient, $m = 1$ and passing through the point $(4, 13)$ is given by:

$$y - 13 = 1(x - 4)$$

So $y = x + 9$

③ $y = \dfrac{2}{3}x^3 + \dfrac{1}{2}x^2 - 6x$

$$\frac{dy}{dx} = 2x^2 + x - 6 = (2x - 3)(x + 2)$$

At the stationary points $\dfrac{dy}{dx} = 0$

$(2x - 3)(x + 2) = 0$

Solving gives $x = \dfrac{3}{2}$ or -2

Substituting $x = \dfrac{3}{2}$ into the equation of the curve to find the y-coordinate gives

$$y = \frac{2}{3}\left(\frac{3}{2}\right)^3 + \frac{1}{2}\left(\frac{3}{2}\right)^2 - 6\left(\frac{3}{2}\right) = \frac{9}{4} + \frac{9}{8} - 9 = -5\frac{5}{8}$$

Substituting $x = -2$ into the equation of the curve to find the y-coordinate gives

$$y = \frac{2}{3}(-2)^3 + \frac{1}{2}(-2)^2 - 6(-2) = -\frac{16}{3} + 2 + 12 = 8\frac{2}{3}$$

Finding the second derivative

$$\frac{d^2y}{dx^2} = 4x + 1$$

When $x = \dfrac{3}{2}$, $\dfrac{d^2y}{dx^2} = 7$. The positive value shows that $\left(\dfrac{3}{2}, -5\dfrac{5}{8}\right)$ is a minimum point.

When $x = -2$, $\dfrac{d^2y}{dx^2} = -7$. The negative value shows that $\left(-2, -8\dfrac{2}{3}\right)$ is a maximum point.

④ $f(x) = \sqrt{x^3} + 2x + 5$

Writing the function in index form gives

$$f(x) = x^{\frac{3}{2}} + 2x + 5$$

Differentiating the function we obtain

$$f'(x) = \frac{3}{2}x^{\frac{1}{2}} + 2$$

> Turn any negative and fractional indices back to reciprocals and roots, etc., before substituting numbers in for x.

$$f'(x) = \frac{3}{2}\sqrt{x} + 2$$

When $x = 4$, $f'(x) = \dfrac{3}{2}\sqrt{4} + 2 = 3 + 2 = 5$

This is a positive gradient so $f(x)$ is an increasing function at $x = 4$.

⑤ (a)

x

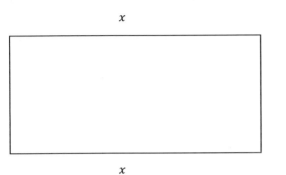

x

If length $= x$, then width $= \dfrac{100 - 2x}{2} = 50 - x$

Area of pen, $A = x(50 - x) = 50x - x^2$

Differentiating the expression for this area gives, $\dfrac{\mathrm{d}A}{\mathrm{d}x} = 50 - 2x$

Maximum value occurs when $\dfrac{\mathrm{d}A}{\mathrm{d}x} = 0$

Hence $50 - 2x = 0$ so $x = 25$ cm

So length $= 25$ cm and width $= 50 - x = 50 - 25 = 25$ cm

There is only one value of x so it must be the maximum value.

We can check that it is the maximum by finding the second derivative.

$\dfrac{\mathrm{d}^2 A}{\mathrm{d}x^2} = -2$ (The negative value proves that the only value for x is a maximum.)

So, the length and width are the same at 25 cm and the pen will need to be square.

(b) Maximum area $= x^2 = 25^2 = 625$ cm^2

Core 2 Pure Mathematics 2

1 Sequences, arithmetic series and geometric series

① $a = 4$ and $d = 6$

$S_n = \dfrac{n}{2}\big[2a + (n-1)d\big]$

$S_n = \dfrac{n}{2}\big[2 \times 4 + (n-1)6\big]$

$$S_n = \frac{n}{2}[8 + 6n - 6]$$

$$S_n = \frac{n}{2}(6n + 2)$$
$$S_n = n(3n + 1)$$

② $S_n = \frac{n}{2}\left[2a + (n-1)d\right]$

$$S_7 = \frac{7}{2}\left[2a + (7-1)d\right]$$

$$182 = \frac{7}{2}[2a + 6d]$$
$$a + 3d = 26 \tag{1}$$
n^{th} term $= a + (n-1)d$

5th term $= a + 4d$ and 7th term $= a + 6d$

$$a + 4d + a + 6d = 80$$

$$2a + 10d = 80$$

Dividing this equation by 2 gives

$$a + 5d = 40 \tag{2}$$

Subtracting equation (1) from equation (2) gives

$$2d = 14$$

$$d = 7$$

Substituting $d = 7$ into equation (1) gives

$$26 = a + 21$$

$$a = 5$$

③ $a + ar = 2.7 \tag{1}$

$$S_\infty = \frac{a}{1-r}$$

$$3.6 = \frac{a}{1-r}$$

$$a = 3.6(1 - r)$$

Substituting this into equation (1) gives

$$3.6(1 - r) + 3.6(1 - r)r = 2.7$$

$$3.6 - 3.6r + 3.6\,r - 3.6r^2 = 2.7$$

$$3.6r^2 = 0.9$$

$$r^2 = \frac{1}{4}$$

$$r = \pm\frac{1}{2}$$

As r must be positive, $r = \dfrac{1}{2}$

$a = 3.6(1 - r)$

$a = 3.6\left(1 - \dfrac{1}{2}\right)$

$a = 1.8$

2 Logarithms and their uses

① $\log_2 36 - 2\log_2 15 + \log_2 100$

$= \log_2 36 - \log_2 15^2 + \log_2 100$

$= \log_2 36 - \log_2 225 + \log_2 100$

$= \log_2\left(\dfrac{36 \times 100}{225}\right)$

$= \log_2 16$

② $\log_{27} x = \dfrac{2}{3}$

$x = 27^{\frac{2}{3}}$

$x = \sqrt[3]{27^2}$

$x = 3^2$

$x = 9$

> The squaring or cube rooting can be done in either order. It is easier, however, to cube root first and then square.

③ $3^x = 2$

Taking logs to base 10 of both sides

$\log 3^x = \log 2$

$x \log 3 = \log 2$

$x = \dfrac{\log 2}{\log 3}$

$x = 0.63$ (2 d.p.)

④ $\dfrac{1}{2}\log_a 36 - 2\log_a 6 + \log_a 4$

$= \log_a 36^{\frac{1}{2}} - \log_a 6^2 + \log_a 4$

$= \log_a 6 - \log_a 36 + \log_a 4$

$= \log_a\left(\dfrac{6 \times 4}{36}\right)$

$= \log_a\left(\dfrac{2}{3}\right)$

⑤ $\log_a(6x^2 + 5) - \log_a x = \log_a 17$

$\log_a\left(\dfrac{6x^2+5}{x}\right) = \log_a 17$

$\dfrac{6x^2+5}{x} = 17$

$6x^2 + 5 = 17x$

$6x^2 - 17x + 5 = 0$

$(3x - 1)(2x - 5) = 0$

Giving $x = \dfrac{1}{3}$ or $x = \dfrac{5}{2}$

3 Coordinate geometry of the circle

① (a) Comparing the equation $x^2 + y^2 - 8x - 6y = 0$ with the equation

Alternatively, you could use the method of completing the square here.

$x^2 + y^2 + 2gx + 2fy + c = 0$ we can see $g = -4$, $f = -3$ and $c = 0$.

Centre A has coordinates $(-g, -f) = (4, 3)$

Radius $= \sqrt{g^2 + f^2 - c} = \sqrt{(-4)^2 + (-3)^2 - 0} = \sqrt{25} = 5$

(b) Rearranging the equation of the straight line for y gives

$y = -2x - 4$

Substituting y into the equation of the circle gives:

$x^2 + (-2x - 4)^2 - 8x - 6(-2x - 4) = 0$

$x^2 + 4x^2 + 16x + 16 - 8x + 12x + 24 = 0$

$5x^2 + 20x + 40 = 0$

$x^2 + 4x + 8 = 0$

Discriminant $= b^2 - 4ac = 16 - 4 \times 1 \times 8 = 16 - 32 = -16$

As $b^2 - 4ac < 0$ there are no real roots so the circle and line do not intersect.

② (a) Comparing the equation $x^2 + y^2 - 4x + 6y - 3 = 0$ with the equation

$x^2 + y^2 + 2gx + 2fy + c = 0$ we can see $g = -2$, $f = 3$ and $c = -3$.

Centre A has coordinates $(-g, -f) = (2, -3)$

Radius $= \sqrt{g^2 + f^2 - c} = \sqrt{(-2)^2 + (3)^2 - (-3)} = \sqrt{16} = 4$

Alternatively, you could use the method of completing the square here.

(b) If point P (2, 1) lies on the circle its coordinates will satisfy the equation of the circle

$x^2 + y^2 - 4x + 6y - 3 = 0$

$x^2 + y^2 - 4x + 6y - 3 = (2)^2 + (1)^2 - 4(2) + 6(1) - 3 = 4 + 1 - 8 + 6 - 3 = 0$

Both sides of the equation equal zero so $P(2, 1)$ lies on the circle.

③ (a) Equation of the circle is:

$$(x-a)^2 + (y-b)^2 = r^2$$

$$(x-2)^2 + (y-3)^2 = 25$$

$$x^2 - 4x + 4 + y^2 - 6y + 9 = 25$$

$$x^2 + y^2 - 4x - 6y - 12 = 0$$

(b) Gradient of line joining the centre $A(2, 3)$ to $P(5, 7)$

$$= \frac{7-3}{5-2} = \frac{4}{3}$$

Gradient of tangent $= -\dfrac{3}{4}$

> AP is a radius of the circle and will make an angle of 90° to the tangent at point P.

Equation of tangent is

$$y - 7 = -\frac{3}{4}(x - 5)$$

$$4y - 28 = -3x + 15$$

$$3x + 4y - 43 = 0$$

4 Trigonometry

① (a) Area $= \dfrac{1}{2}bc\sin A = \dfrac{1}{2} \times 12 \times 8 \times \sin 150° = 24\,\text{cm}^2$

(b) Using the cosine rule

$$a^2 = b^2 + c^2 - 2bc \cos A$$

$$= 12^2 + 8^2 - 2 \times 12 \times 8 \cos 150°$$

$$= 374.2769$$

$$a = 19.3462$$

$$a = 19.3\,\text{cm (to one decimal place)}$$

> Remember not to round off answers to the required number of decimal places until the final answer.

② (a) $(0, 0), (\pi, 0), (2\pi, 0), (3\pi, 0), (4\pi, 0)$

(b) $\left(\dfrac{\pi}{2}, 1\right)\left(\dfrac{3\pi}{2}, -1\right)\left(\dfrac{5\pi}{2}, 1\right)\left(\dfrac{7\pi}{2}, -1\right)$

③

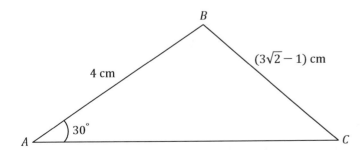

Using the sine rule

$$\frac{\sin C}{c} = \frac{\sin A}{a}$$

> No diagram is given in the question, so draw your own and add the information. You can now see the pairing of the sides and angles which points to the need to use the sine rule.

$$\frac{\sin A\hat{C}B}{4} = \frac{\sin 30°}{3\sqrt{2} - 1}$$

$$\sin A\hat{C}B = \frac{4\sin 30°}{3\sqrt{2} - 1}$$

$$\sin A\hat{C}B = \frac{2}{3\sqrt{2} - 1} = \frac{2(3\sqrt{2} + 1)}{(3\sqrt{2} - 1)(3\sqrt{2} + 1)}$$

> Remove the surd from the denominator by multiplying the numerator and denominator by the conjugate of the denominator (i.e. $(3\sqrt{2} + 1)$).

$$\sin A\hat{C}B = \frac{6\sqrt{2} + 2}{18 - 1} = \frac{2 + 6\sqrt{2}}{17}$$

④ $3\sin^2 \theta = 5 - 5\cos\theta$

$3(1 - \cos^2\theta) = 5 - 5\cos\theta$

$3 - 3\cos^2\theta = 5 - 5\cos\theta$

$3\cos^2\theta - 5\cos\theta + 2 = 0$

$(\cos\theta - 1)(3\cos\theta - 2) = 0$

$\cos\theta = 1$ or $\cos\theta = \dfrac{2}{3}$

> Use the trig identity $\cos^2\theta + \sin^2\theta = 1$ to create a quadratic equation in $\cos\theta$ which can then be solved.

$\theta = \cos^{-1}(1) = 0°, 360°$ or $\theta = \cos^{-1}\left(\dfrac{2}{3}\right) = 48.2°, (360 - 48.2)°$

Hence $\theta = 0°, 48.2°, 311.8°$ or $360°$

5 Integration

① $\displaystyle\int\left(4x^{\frac{1}{3}} - \frac{2}{\sqrt[3]{x}}\right) = \int\left(4x^{\frac{1}{3}} - 2x^{-\frac{1}{3}}\right)dx$

$$= \frac{4x^{\frac{4}{3}}}{\frac{4}{3}} - \frac{2x^{\frac{2}{3}}}{\frac{2}{3}} + c$$

$$= 3x^{\frac{4}{3}} - 3x^{\frac{2}{3}} + c$$

② $\displaystyle\int\left(\sqrt[3]{x} + \frac{1}{x^4}\right)dx = \int\left(x^{\frac{1}{3}} + x^{-4}\right)dx$

$$= \frac{x^{\frac{4}{3}}}{\frac{4}{3}} + \frac{x^{-3}}{-3} + c$$

$$= \frac{3}{4}x^{\frac{4}{3}} - \frac{x^{-3}}{3} + c$$

③ $\int\left(\dfrac{4}{x^3}-6x^{\frac{1}{5}}\right)dx = \int\left(4x^{-3}-6x^{\frac{1}{5}}\right)dx$

$$= \dfrac{4x^{-2}}{-2}-\dfrac{6x^{\frac{6}{5}}}{\dfrac{6}{5}}+c$$

$$= -2x^{-2}-5x^{\frac{6}{5}}+c$$

④ $\int\left(\dfrac{2}{\sqrt{x}}-x^{\frac{3}{2}}\right)dx = \int\left(2x^{-\frac{1}{2}}-x^{\frac{3}{2}}\right)dx$

$$= \dfrac{2x^{\frac{1}{2}}}{\dfrac{1}{2}}-\dfrac{x^{\frac{5}{2}}}{\dfrac{5}{2}}+c$$

$$= 4x^{\frac{1}{2}}-\dfrac{2}{5}x^{\frac{5}{2}}+c$$

⑤ $\int_0^4\left(x^{-\frac{1}{2}}+2x\right)dx = \left[2x^{\frac{1}{2}}+x^2\right]_0^4 = \left[2\sqrt{x}+x^2\right]_0^4 = \left[(4+16)-(0)\right]=20$

⑥ $\int_a^b y\,dx \approx \dfrac{1}{2}h\left\{(y_0+y_n)+2(y_1+y_2+\ldots+y_{n-1})\right\}$

$\int_0^4\dfrac{1}{1+\sqrt{x}}\,dx \approx \dfrac{1}{2}h\left\{(y_0+y_n)+2(y_1+y_2+\ldots+y_{n-1})\right\}$

$h = \dfrac{b-a}{n} = \dfrac{4-0}{4} = 1$

When

$x = 0,\ y_0 = \dfrac{1}{1+\sqrt{0}} = 1$

$x = 1,\ y_1 = \dfrac{1}{1+\sqrt{1}} = 0.5$

$x = 2,\ y_2 = \dfrac{1}{1+\sqrt{2}} = 0.41421$

$x = 3,\ y_3 = \dfrac{1}{1+\sqrt{3}} = 0.36603$

$x = 4,\ y_4 = \dfrac{1}{1+\sqrt{4}} = 0.33333$

$\int_0^4\dfrac{1}{1+\sqrt{x}}\,dx \approx \dfrac{1}{2}\times 1\left\{(1+0.33333+2(0.5+0.41421+0.36603)\right\}$

$\approx 1.94691 \approx 1.947$ to 3 decimal places